Praise for
On Digital Advocacy

"The more time we spend online, the easier it can feel to slide into binary ways of thinking: indoors or outside, burn out or sign off, advocate or give up. In the face of all *that*, Boué's *On Digital Advocacy* is a refreshing invitation to breathe and get curious about the things that matter most to us. And along the way, Boué generously grants us the tools we need to make a real difference—all delivered with her trademark gusto and grace."

—Gale Straub, author of *She Explores:
Stories of Life-Changing Adventures
on the Road and in the Wild*

"For every outdoor lover asking, 'How can I make a difference?' Katie Boué's *On Digital Advocacy* offers a clear guide for how to make an impact. The world of advocacy is changing quickly, and throughout the book, Boué's powerful voice and expertise guide you through strategies and examples for modern activism and organizing. Whether you're an individual or an organization, learn how to create impact that matters using Boué's unique combination of storytelling and case studies that are sure to inspire. Reading this book will support you in your journey to making the world a better place, whatever your cause."

—Caroline Gleich, professional skier
and climate activist

"Boué's book comes at a critical time when the worlds of advocacy and community are colliding with individuals' online presence. I think it's a must-read for folks interested in digital advocacy. But really, it's a crucial conversation for anyone working to create a true impact between their digital profiles and neighborhood community."

—Heather Balogh Rochfort,
award-winning adventure writer

"I looked up Outdoor Advocacy in the dictionary. It said: Katie Boué. From being a creative force behind signature campaigns (Vote the Outdoors) to demonstrating how advocacy can be part of a lifestyle (Outdoor Advocacy Project) and more (oh, so much more!), Katie has been a beacon to me when it comes to advocacy work in and for the outdoors. And now she has it in book form, providing a how-to and why-to grounded in her depth of expertise and lived experience. But one thing that Katie deeply embodies and which I respect and admire is the growth-oriented process that is crucial to this work and that keeps our humanity centered as we fight for people and planet in ways that do not destructively consume us. Because we have enough 'doom and gloom,' we need more 'do and bloom' that activates individual action in relation to our communities so we can thrive. And Katie offers us that. So do yourself a favor and read *On Digital Advocacy: Protecting the Planet While Preserving Our Humanity. Y gracias*, Katie, for your *palabras*."

—José Gonzalez, Founder
at Latino Outdoors,
Educator, Creative, Facilitator

ON DIGITAL
ADVOCACY

The Speaker's Corner Books Series

Speaker's Corner Books is a series of book-length essays on important social, political, scientific, and cultural topics. Originally created in 2005, the series is inspired by Speakers' Corner in London's Hyde Park, a bastion of free speech and expression. The series is influenced by the legacy of Michel de Montaigne, who first raised the essay to an art form. The essence of the series is to promote lifelong learning, introducing the public to interesting and important topics through short essays, while highlighting the voices of contributors who have something significant and important to share with the world.

ON DIGITAL ADVOCACY

*Saving the Planet While
Preserving Our Humanity*

Katie Boué

Fulcrum Publishing
Wheat Ridge, Colorado

Library of Congress Cataloging-in-Publication Data
is available
ISBN 978-1-68275-345-3
eISBN 978-1-68275-370-5

Printed in the United States
0 9 8 7 6 5 4 3 2 1

Cover art and design by Kateri Kramer
Interior illustrations by Kateri Kramer
Map courtesy of the author

Unless otherwise noted, all websites cited were current as of
the initial edition of this book.

Fulcrum Publishing
3970 Youngfield Street
Wheat Ridge, Colorado 80033
(800) 992-2908 • (303) 277-1623
www.fulcrumbooks.com

To the Boué family,
for always supporting
your queer little Eliza Thornberry.
And to myself in 1996,
who wanted to write a book
when she grew up.
You did it, Bean.

"Knowing that you love the earth changes you,
activates you to defend and protect and celebrate.
But when you feel that the earth loves you in return,
that feeling transforms the relationship from
a one-way street into a sacred bond."

—Robin Wall Kimmerer

FOREWORD

by Amelia Howe

AMELIA

VOTE

During my first D.C. fly-in lobby event as a policy associate with the American Alpine Club, I was in a state. Events like this are exciting because they offer the opportunity to bring outdoor advocates together from across the county to meet with their lawmakers and speak up on behalf of the issues that matter most to their communities. Not only are these events a good way for lawmakers to hear from their constituents directly, but they can be empowering for the individuals who are making the asks and often leave people with a desire to bring their advocacy skills back to their communities.

In preparation for the event, I spent weeks managing travel logistics for advocates, creating in-depth fact sheets surrounding the issues we would discuss during our lobbying meetings, and scheduling meetings with Senate staffers, congressional chiefs of staff—and occasionally, if we got lucky, even a senator or congressperson themselves. My brain was running on caffeine and spreadsheets, and I felt like an imposter syndrome–stricken fish out of water in my J. Crew outlet blazer.

On the morning of our first lobbying day in the Capitol, I walked into breakfast to meet "team Utah" to prepare for our meeting with Congressman John Curtis. I was equal parts stoked and terrified. Suddenly, while ordering a coffee, a glasses-clad, familiar-looking woman tapped me on the shoulder and asked if I remembered what the heck the acronym for "SOAR" in "SOAR Act" stood for. I recognized her, likely from an Alder Apparel advertisement or through her internet-famous dog named Spaghetti. At that moment, I had no idea that while standing there in my navy blazer and two-inch heels I had just met my long-lost advocacy soulmate and best friend.

Katie met with Congressman Curtis before, and I was in awe of the ways she wooed him with compliments of his eclectic sock collection and references to the time she took him bouldering in Joe's Valley. It was clear that she had a knack for communicating and finding ways to be memorable. That day I learned lobbying is one part data, one part ask, and two parts relationship building. The personal connections we make with people are often the biggest leverage point we have when it is time to make a request. Luckily for me, this particular connection not only led to a meaningful conversation with a congressman but also a lifelong friendship with my teammate Katie.

Our team crushed that meeting. Congressman Curtis cosponsored the Simplifying Outdoor Access to Recreation (SOAR) Act in 2019 and became a big champion of the bill. A quick, momentary win! Alas, due to the inherently political nature of this work, the SOAR Act, as of 2023, still had not been signed into law.

Recently, the organization I work with celebrated a conservation win that took more than twenty years to accomplish. Yes, you read that right—twenty years. After two decades of working with a coalition of local tribal members; recreationists; state, local, and federal governments; and nonprofits, the hydroelectric dams on the Klamath River in southern Oregon and northern California were approved for removal. It is the largest dam removal effort in US history. It means the connection of coldwater fisheries, the protection of cultural and historical fishing traditions for Native Peoples, and more rapids to run for water-sports enthusiasts.

It's easy in the moments of victory to celebrate and immediately move on to the next project. However, when something is twenty years in the making, we should probably slow down and take a moment to revel in it. To speak to the community members who were on the ground in the earliest stages of the project. Ultimately, to learn from the project's tactics and carry the work forward.

In the age of the internet, we expect instant gratification. TikTok videos max out at three minutes, a package can be delivered to your door the same day, and we date by looking at a photo and swiping left and right on our phones. As outdoor advocates and organizers, we are met with an impossible task in this era of instant everything. We must break through the noise, and get people excited about outdoor advocacy, not just for a single Instagram campaign, but for the long haul.

This begs the question: Where do we as individuals who are passionate about the outdoors fit into the policy and advocacy space? I have always been curious about what makes a person stick around to see the end of a twenty-year campaign. Largely, I think the mystery can be solved by a carefully formulated equation of education, empowerment, and trust on the part of the organizers. In this book, Katie reveals a tool we worked on together in an attempt to crack the code of advocate retention. There will also be examples of different campaign tactics that can be useful for keeping your local community and network engaged.

As I work in the policy and advocacy space, it is evident that, as outdoorists, we have a huge advantage in this work.

The stories we have to share are directly tied to place and passion—they're memorable. Often, our stories bridge the gap between community and environment. Depending on who we are trying to sway, we can use bold statistics that point to our economic impact as the outdoor industry, or evoke a powerful memory an individual has in an outdoor space they love in their own community. You as an individual can leverage your story, or the story of your community, for change. When finding your place in the movement, the first step to take is understanding the power behind making your voice heard. Civic engagement and being an active participant in local, state, and national politics doesn't always feel exciting, but it can be when you are in charge of crafting your own narrative, or your story of self.

Working in this space, big wins don't come my way every day. These wins take place after hard-fought, thankless, brutal years of screaming into a perceived void. It is easy to begin to feel like your efforts as an advocate are all for naught, and that the stories you have to share are meaningless when stacked against big-money lobbyists and an ever-widening partisan divide. This book is not here to tell you that those feelings aren't valid. They are. However, hopefully, when walking away from this you will feel empowered by the fact that when communities rise up around a shared objective, they hold meaningful power. Our charge as advocates is to seek out the people in our communities who light our fires, challenge us, and lift us up, and from there work together to combat burn-out in order to keep coming back and advocating for our collective needs.

In the world of outdoor policy, there is so much reaction. There is a lease sale announced on top of a classic mountain bike trail, we have to stop it! Wildfires are getting worse and worse every year, we need the agencies to hire more wildland firefighters! The US Forest Service budget was cut, time to write our senators condemning them for letting this occur! Imagine a world in which every outdoorist is armed with knowledge and empowered to act on issues at any given moment due to the intentional work of a trusted organization or entity. With that level of engagement, outdoorists' voices could elevate to the level of big-money lobbyists, and we could begin to see wide-scale environmental and conservation improvements that withstand time and administrative shifts. For this reason, it is so critical that we are proactive in activating and educating our community members.

This work, like so much else in life, is messy. Rarely are the solutions we advocate for at the start of a campaign the ones we end up with when it's done. Policy work requires strategy, yes, but it also requires us to think quickly and be amenable to change. If we go into a coalition meeting thinking we have all the answers, we will never find solutions that last for the long haul. The most effective outdoor advocates and policy wonks I know are those who see the solution as one that can be altered, expanded, and changed over time.

Compromise isn't easy, but it is required. Currently, I live in Wyoming and along with countless others, have been thinking a lot about the impact of the massive federal bills that have recently passed, funding future environmental movements. How is the money in these bills going to directly impact my state, and most importantly, our communities on the ground? When it comes to solving giant issues like climate change and national energy transitions, it can be rather overwhelming to see where we fit in. Luckily, as individuals, we do not have to figure it all out ourselves. We can take a puzzle-piece approach, find the levers of change we hold, and organize alongside our community to accomplish shared goals.

Here in Wyoming, a major piece of the puzzle being considered is how an energy transition should be addressed. This state is the second-largest energy producer in the country. Oil and gas from Wyoming are likely powering your next big outdoor adventure road trip, keeping your home warm during the winter, or fueling your flight to go backpacking abroad. As we begin to fund the energy transition in this country, western states like Wyoming are also going to see increased exploration for critical and rare earth minerals. This means that in a clean energy future, the resources below our surface may no longer be needed for a gasoline-powered car, but newly mined critical minerals may be used to create the battery that lives in an electric vehicle. This issue highlights the nuance of policy and advocacy work. Energy has been the narrative behind a healthy and vibrant livelihood in Wyoming for decades, and that is not going to change overnight. As advocates, large issues like this need to be handled delicately and with humility. Our role is to actively listen to a wide variety of community members' concerns and to continue to show up and use our voices in support of moving toward a more sustainable future. This could look like getting a friend group together to attend a city council meeting, hosting a letter-writing party, or encouraging a colleague to run for the state legislature. Finding solutions that are lasting and impactful simply begins with engagement, not only federally but in your own backyard.

Similarly, a question that we have to be asking as we make changes to combat the climate crisis is what will happen to the individuals who work in the energy sector when we make these large-scale grid transitions? We have seen the impacts from our transition away from coal. We can choose to quit the resource, but we can't quit the people who power our grid. We can't quit the rural communities that have been built around natural resource extraction. It is our responsibility as advocates to find solutions to ensure that these transitions are just, equitable, and being led by the folks who are going to see the biggest change in their actual lives. We, as citizens, need to hold our local, state, and federal governments accountable to ensure communities are not just supported but also are empowered along the way.

People tell you all the time that your work shouldn't be your everything. Messages reminding you that work-life balance is important and learning how to say no is a gift are being posted by LinkedIn influencers daily. In 2019 my response to said influencers would have been, "I hear you, but have you ever worked in outdoor advocacy?"

Things can get complicated when your work, passion, and hobbies all become intertwined with your advocacy efforts. Suddenly, it is hard to create and uphold healthy boundaries. I certainly fall victim to this. However, let this book teach you that your voice and unique story matter and hold power.

Heed Katie's advice on allowing yourself to rest. The outdoor advocacy movement needs people to stick around for the twenty-year efforts, and, as anyone who has run an endurance race understands, this requires intentional fuel, rest, and a crew to motivate you.

Finally, celebrate the little wins. Knowing that the big wins are few and far between, allow yourself to feel joy with each moment that leads you closer to your objective. No accomplishment is too small to celebrate.

LETTER TO OUTDOORISTS (THAT'S YOU)

I believe every outdoorist (any human, really) has
a responsibility to take care of the planet.

Dear Outdoorist,

Whether we hike, bike, camp, climb, hunt, ride, paddle, paint, garden—whatever way we get out and enjoy
nature—we leave an impact on the outdoors every time
we step outside. Every step our boots take down dusty
trails, every bolt we clip draws into, every time we cruise
down a dirt road, we leave an impact.

There's a trail winding through my favorite place on
earth, the San Rafael Swell, cutting straight through
miles and miles of cryptobiotic soil fields. Crypto, as it's

called, is a living soil crust composed of organisms like algae, fungi, and cyanobacteria. It holds the entire desert together, provides arid plants with a rare home in which to root, and takes decades to form. Black earthen castles spread across the landscape like tiny kingdoms, and a single footstep can crush a hundred years of careful growth in an instant.

The Goodwater Rim Trail slices through fifteen miles of the soil crust's home like a clean scar, either side of a sandy single track flanked by blackened biotic expanses. The building of this trail, and even our daily use of it, means inevitable harm to the cryptobiotic soil. But would I have become such an advocate for this organism had I not had access to a trail leading me to it? What is the cost of creating an advocate? Is it worth it?

I think so. Though I certainly busted my fair share of crust before I knew better, learning about the sacred nature of cryptobiotic soil changed and deepened my relationship with this landscape I have come to be an advocate for. I began to consider the many layers of this ecosystem, beyond the scope of my own recreation.

I began to witness.

Miniature castles of black, white, and pink crypto towering along sandstone hillsides, soil sentinels holding the land together through wind and rain. A bizarre cocktail of algae, moss, lichen, bacteria, and fungi—all swirled together to make the desert a sustainable home. The soils became a friend to me. If you asked a friend of mine to write a list of the things I love, there's a strong chance they'd scribble "crypto" toward the top alongside my dog, Spaghetti, and my gardens.

Maybe you know what I mean. Maybe you have a natural companion you've grown to recognize and adore. Perhaps it's juicy banana slugs creeping through the redwoods or Spanish moss draping itself down cypress trees in the South. Whatever it is, that thing pulling your heartstrings and tugging you to *do* something is worth putting your boots on the trail for.

The tricky thing is: regardless of our love for these places, we're still having an impact on them. This means we have a responsibility, a sacred duty, to give back to them.

It's easy for us to point fingers at extractive industries like oil and gas.

"That's bad for the planet," we confidently accuse.

Drilling on public lands? An obvious negative impact on the land and a resounding "no thanks!" from us. Outdoorists, however, aren't absolved from our own impacts just because we hold beloved affection for these natural spaces. The scales of impact are different, of course, and nuance is a necessary companion to every aspect of this conversation.

We have an impact on the land every time we step outside. That matters.

My first lessons as a student of earth's curriculum came before my ability to keep memories, but the idea that nature and all its creations should be respected and loved was a constant in my household. My mother wrapped earthworms around my fingers while I played in the sticky dirt in Miami. My father taught me how to save frogs from the pool filter, and we'd walk down to the canal to gently release them. I spent summers at a Girl Scout camp covered with potato vines where bugs and birds always had the right of way. I was taught early that we take only memories and leave only footprints. But leaving footprints is leaving something.

Outdoor recreation inherently involves impact on the land. Whether we feel at ease justifying it or not, it is the simple truth. Extractive industries aren't the only ones taking from the land. Climbers drill anchors into cliff faces, hikers build zigzagging trails, paddlers shape riverbanks for take-outs, campers affect vegetation every time we set up our tents.

This is not to demonize outdoor recreationalists or to conflate our individual impacts with the detrimental scale of the oil and gas industry—but we must acknowledge them nonetheless. I believe that once we take honest ownership of our impact on the outdoors, we're more empowered to do something effective about it.

If we are to be moved to protect these places, we must first fall in love with the land. We must connect with it. The land must become our holy place. And for more people to sustainably experience these places and fall in love with them, development of infrastructure must happen.

We need more outdoor advocates, yes, but those budding advocates will need parking, bathrooms, visitor centers,
and safe ways to visit. It's a messy and imperfect system, but
problem-solving is a core pastime for adventurers, right?

A key challenge to these notions is this: individuals
cannot be held responsible for solving all the ailments
of this mistreated planet. The brunt of the climate crisis
should be on the shoulders of industry and corporations.

You will not save the planet.

I won't either.

That's okay.

Release yourself from that burden immediately.

The onus to radically invest in the heaviest labor of
solving the climate crisis is on corporations (looking at
you, outdoor industry), period. It's the people, communities, and individuals that will demand change from that
corporate level. After a decade of working in the outdoor
recreation industry, I've grown weary of the idea that
change will be sparked from the top down.

Change will be demanded by the people who will push the
corporations (run by individual people like us, mind you) to
enact the scale of change needed to stop this planet from erupt-
ing into a fiery hellscape. It shines clearly to me that positive
impact will radiate from the bottom up. If we want to change
the corporate harm, the people must make our demands.

Conversations around climate solutions and environmental issues are often framed with a simple dichotomy:

corporate versus individual. Goliath versus David. The scene is set: a pile of white dudes in suits, frothing at the mouth, ready to obliterate you, a lone individual wielding nothing but a reusable straw and a recycled hemp shirt.

Honestly, I'd be unlikely to engage in that battle too.

But the scene is staged unfairly. In the real world, it's not one sole martyr against The Man—it's all of us, together. A pack of corporate schmucks rabid with greed is intimidating, yes, but so is a herd of unshowered hikers who just heard their favorite trail might turn into housing sprawl. If you're getting hands-on with the *On Digital Advocacy* workbook (you should be), there's a journal prompt on page 10 and a case study about one such situation in the Southeast, where Georgia climbers rallied together to save the Boat Rock bouldering area from being bulldozed for a suburban housing development.

When I ruminate on our collective and constantly missed opportunities to galvanize around outdoorist advocacy, outdoor recreation economy statistics always come to mind. The average backpacker likely doesn't know them—yet—but every corporate outdoor industry professional does. I learned the numbers while cutting my teeth as a young social media coordinator for Outdoor Industry Association (OIA), the trade group representing retailers, manufacturers, and organizations in its namesake sector. The details have changed over the years, especially since the COVID-19 pandemic began, but when the statistics were first released in 2017, they read: the outdoor industry generates $887 billion in consumer spending annually and creates 6.1 million American jobs across the country (source: https://outdoorindustry.org/wp-content/uploads/2017/04/OIA_RecEconomy_FINAL_Single.pdf#:~:text=This%20%24887%20billion%20in%20annual,state%20and%20local%20tax%20revenue.&text=The%20livelihoods%20of%207.6%20million%20Americans%20depend%20on%20outdoor%20recreation).

That's a lot.

Every time I met with federal representatives on Capitol Hill, listened to leadership speak during events, or prepped a press release for journalists during that era, the $887 billion industry data was cited. It was our industry's darling party trick, the only ammunition we had to back up our claims that the outdoor sector should be taken seriously in politics and economics.

Since the first day I started working with the outdoor industry, and long before my time, pushing for economic recognition was a key priority for the sector. It was a squeaky wheel, and we wanted grease.

It worked. On December 8, 2016, President Barack Obama signed the Outdoor Recreation Jobs and Economic Impact Act into law—legislation that organizations like OIA lobbied hard in support of. The data was released by the Bureau of Economic Analysis (BEA) in 2018 and indicated that the outdoor recreation industry comprised a whopping 2.2 percent of the entire US Gross Domestic Product (source: https://www.bea.gov/news/blog/2018-09-20/outdoor-recreation-grew-faster-us-economy-2016; https://www.bea.gov/news/2018/outdoor-recreation-satellite-account-updated-statistics-2012-2016).

If those numbers fail to impress you, consider this: the gas, oil, and mining extraction industries only amounted to 1.4 percent in the same study.

Yeah.

The report conventionally defined the outdoor recreation industry as "all recreational activities undertaken for pleasure that generally involve some level of intentional physical exertion and occur in nature-based environments outdoors." It also offered a broader definition that included "all recreational activities undertaken for pleasure that occur outdoors."

Because we all know taking a nap under a pine tree counts as outdoor recreation too.

Here's the thing about those numbers: they're used by the *industry*, but they're measuring the power of the *community*. The BEA numbers clocked annual consumer spending at $412 billion in 2016. This measured every dollar spent when you packed up the car for an outdoor adventure. Every piece of camping gear, every campsite reservation, your dirt bike, going to outdoor concerts, the gas in your tank to get you there—it all counts as part of the impact that the outdoor community has on nationwide GDP.

As for the job numbers? They're an indicator of how many individual built-in outdoor advocates our industry has. It's part of why I stick to this industry despite how much it frustrates me sometimes.

We're an economic powerhouse that is truly values aligned.

Someone who works for a hiking boot company is likely to be a hiker themselves—and who better suited to take up the duty of protecting hiking trails?

These numbers legitimized our industry. Lawmakers, public lands managers, and the folks in charge of what's happening to the outdoors now had established irrefutably legitimate data to demonstrate the positive influence of outdoor recreation on our economies. Making this country better by investing in the outdoors was no longer a nice idea; it was a proven fact.

During my first visits to lobby on behalf of the outdoor industry in Washington, D.C., I was a small fish in a big pond—but hearing mentors rattle off state-specific numbers about the size and strength of the outdoor sector, I was emboldened. Soon, I was parading down the halls of Capitol Hill to meetings with senators and representatives, citing outdoor economic stats with confident authority. Lawmakers listened, and the outdoor contingency grew more powerful with every interaction.

I eventually started to question my role in the outdoor space because of this data. Numerous outdoor "industry" political stances were at odds with rising outdoor "community" values. It was frustrating to watch the *industry* leaning on data that represented the purchasing power of the *community*. The industry had talking heads, well-spoken businesspeople, and long-established political relationships. The community seemed to be just a bunch of riled-up advocates without a lick of formal lobbying skills.

I wanted to be with the community.

Understanding the skills and connections gleaned during my time working on the industry side of things, I wanted to distribute my knowledge to serve the community. I knew I couldn't do that while still obligated to the policy agenda of the Outdoor Industry Association, so I ended my contract.

Released from the limitations of the industry and fueled by the audacity my ego had built up over years of positive feedback on my communication style from outdoor mentors, I set out to create a space where advocacy could become more accessible and more actionable to our community.

So I founded the Outdoor Advocacy Project.

But before all of that, I was just a Florida kid living in a rusty van.

VAN LIFE (2012–2014)

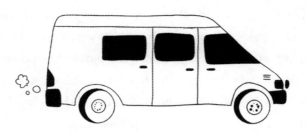

In 2012, I was twenty-four years old and decided to buy a big yellow van in pursuit of climbing. I had fallen in love with bouldering and rock climbing at Tallahassee Rock Gym while attending Florida State University—and any Florida climber knows you have to commit to road trips if you want to pull on real rock and not just plastic in gyms. From The Sunshine State panhandle, it was a minimum of six hours in the car to the nearest climbing area. My college climging friends and I made the commute toward sandstone every weekend we could.

After graduation, I took a six-week cross-country road trip to climb granite in Yosemite, sandstone in Arches, choss in Garden of the Gods, and more roadside boulders than I could count.

That brief taste of the road wasn't enough, though. I wanted to travel full-time and see everything public lands in the United States had to offer.

It was during the very beginning of the Van Life era that I poured my entire life savings into an already-rusting 2004 Dodge Sprinter. My dad and I booked one-way flights to Tampa, and I learned how to haggle with used-car salespeople. After a mess of paperwork, my father secretly recording me giving attitude to the salesman, and a quick lesson on how to drive it, I had my van's keys in hand.

The first time I ever drove the big marigold cargo van was across the Sunshine Skyway Bridge, a four-mile trek across the bay that towers 430 feet above the water—in a severe wind advisory. We could feel the enormous structure moving ever so slightly as we drove up the four-lane beam bridge toward the sea-blue sky.

"Hold on, Kate, I think we're in for a wild ride!" my dad gritted through his teeth while white-knuckling the dashboard.

Turns out, I was.

The hightop van swayed back and forth like a sailboat in rough waters, but we made it back in one piece. I never did get comfortable driving it, though, and would spend much of the next year in the passenger seat.

I spent about $500 on materials to make a rickety living space using discounted house furnishings from IKEA and Home Depot clearance sections. Spoiler alert: my entire kitchen would become unhinged from its mediocre brackets and start sliding around one day while we scrambled across rock crawls way out of our league on the way to Bridger Jack Campground in Utah's Indian Creek. If you decide to build a camper van, don't skimp on construction costs.

I wanted to see the country, explore its landscapes, and Go Out West—whatever that means.

I wanted to write about it.

Having freshly earned my BA in creative writing, I felt deeply inspired by a Traveling Writing workshop in my final semester. I started a blog, joined new social media sites like Twitter and Instagram, and sent emails to climbing brands asking if they'd be interested in exchanging gear for words and photos.

The first sponsorship came from Columbia Sportswear. After finding my profile on Twitter, their social media manager, Adam Buchanan, sent a mysterious package containing a GoPro and a branded letter inviting me to be part of their inaugural ambassador program called Omniten. It was a group of ten everyday adventurers invited to test gear, attend free press trips to places like Costa Rica, and represent the brand. Eventually, this model would become the standard for ambassador programs across the outdoor industry.

To me in that moment, it was a revelation. Adventure, storytelling, the outdoors—this could be what I "do" when I grow up.

For the entirety of 2013, I lived out of Southern Comfort—my then-partner and I named the van after its original calling as a Southern Comfort One-Hour Heating & Air service vehicle. The name also paid homage to the South, where my love of climbing and public lands deepened into a lifelong commitment to advocacy.

It was in the boulder fields and backroads of Tennessee, Georgia, Alabama, and Florida where I learned the philosophies of stewardship, community care, treading lightly, and political activation in the name of protecting the outdoors. I just didn't know that jargon yet. The vocabulary of advocacy empowered me, so I've included a glossary of terms in this book to make the language more accessible to us all.

I learned these virtues through practice, and in reflection of the current virtues-by-Instagram-infographics age, I am grateful to have collected my personal values through hands-on experience rather than through screens.

My sense of responsibility to public lands came through being in relationship with these places.

My rent was paid in campground fees and parks passes. The outdoors was my home, my office, my bathroom, my backyard. I was a transient tenant on America's public lands, and it didn't take long for me to develop a sense of needing to give back to the places I was taking so much from.

I didn't know Robin Wall Kimmerer's concept of being in reciprocity with the land by that term yet, but I knew it by feeling. It was a guttural pull, an instinctual tugging toward balance: I take from the earth, and so I must give back to it.

Those inklings toward stewardship climaxed during my stint in Joe's Valley, Utah. What began as a quick five-day pit stop at an obscure bouldering destination on our way to greater pursuits turned into a three-month stay—and eventually would be part of why I moved to Utah permanently.

The gritty sandstone bouldering lines, the easy camping with like-minded neighbors, the collective sense of pride and wonder to be climbing at such a special place, those Butterfinger donuts—Joe's Valley will forever be my favorite climbing destination on earth.

One day while catching up on wifi and donuts at a gas station/grocery store hybrid called the Food Ranch (if you know, you know), we saw a flyer advertising the town's annual cleanup. Figuring we more than owed the local community some gratitude for our extended stay, our crew of six climbers showed up to Orangeville City Park ready to give back.

The locals took one look at our rancid, disheveled bodies pouring out of a rusty van and immediately had questions.

"So, are y'all here for the free food?"

Excitedly, we replied, "No, ma'am, we're here to help!"

Skeptical, they handed us shovels and put us to work. As we dug a season's worth of dead leaves from street gutters, a cultural exchange of sorts began to transpire.

"We always see y'all walking off into the woods with those mattresses on your backs—what's that all about?"

We explained the sport of bouldering to the locals, gesturing as we described how crash pads helped us stay safe during falls. They told us about the rich coal-fueled history of the town and why they loved rural life in Emery County. We debunked their curiosities about living in vehicles full-time and gushed over how we fell in love with the sandstone, Butterfinger donuts, and slow pace of life in Joe's Valley.

We were wildly different humans, but we shared a deep commonality through that place. Boulderers used the landscape to huck ourselves up rocks and as a safe place to rest our bodies after a long day of climbing. Locals used the land a little more practically—for agriculture, hunting, and fishing. Our activity preferences weren't always aligned, but our reverence for Joe's Valley was the same. The land could bring us together.

Our climber cohort did eat a free lunch in the park that day. But more importantly, that day spent shoulder-to-shoulder cleaning up the town with locals sparked the beginning of a deep mutual relationship between climbers and rural residents.

The next time I drove through Orangeville, there was an enormous vinyl banner flapping in the wind outside of the Food Ranch. It read: "We love climbers!"

Nine years after that city cleanup day, I received an email from the Joe's Valley Festival with the subject line, "Let's give back to Orangeville, Utah." It was an invitation from the Joe's Valley Advocacy Group to meet for the annual Orangeville City Park cleanup. I wonder if they fed the climbers a free lunch afterward.

At the end of my year in the van, I sold Southern Comfort, was promptly cheated on by my partner, moved to Denver on a whim, and ended up in an interview for a part-time social media position with the Outdoor Industry Association (OIA). I was looking for something to move me, to shake me—and that thing was outdoor advocacy.

BECOMING AN ADVOCATE FOR THE OUTDOOR INDUSTRY

It was Halloween 2014 when I first walked into the offices at OIA. Everyone was dressed up in costumes—except me. As I was interviewed by a traffic cone and Max from *Where the Wild Things Are,* I declared that while I understood this was a part-time social media coordinator position, I intended to make it a full-time career.

Years later, my then-boss, Jenn Brunson, would be quoted in a magazine feature naming me one of the twenty-six best and brightest people in the outdoor industry saying, "She's kind of an anomaly. It's rare that you find someone who is so, so passionate about what they do and how they recreate."

I was hired on a ninety-day trial to see if social media was a worthwhile endeavor. My hypothesis was that social media would become the backbone of communication. I was correct. Within a few years, the success of my social media campaigns would lead me to a grassy meadow in Yosemite National Park, live tweeting as President Obama quoted our economic data—but we're not there yet.

Through organic social media strategies, I took OIA's Instagram platform from 124 followers to 24,000 and helped double our website traffic during my tenure there. I leaned heavily on grassroots tactics like personalized outreach, mutual collaboration, amplification, and used social media to build relationships. More importantly, my time at the organization introduced me to policy making and the nitty gritty of outdoor advocacy—and the dark side of going viral.

A lesson I've learned about viral moments on social media is: you never know when they're coming. I smacked face-first into this lesson during my first big event as a social media coordinator.

Outdoor Retailer (OR) is a biannual industry convention that was once the most anticipated convening of the year, and my first trade show with OIA—title sponsor of the event—came in January 2015, just weeks after I started what I was hoping to make a full-time job. It was also my first experience with the flu, which left me green-faced, bedridden, and live tweeting events from my tepid hotel room across from the convention center.

During our signature keynote breakfast, the hired motivational speaker quoted J. K. Rowling in a context so insignificant that I can't remember it now—but as a Harry Potter fan, I was delighted to tweet out the quote and a picture of the slide, tagging the author.

Within minutes, a notification buzzed on my phone.

She had responded.

I was elated.

My fingertips rushed to slide open Twitter.

"I have never said that in my life. Never. . . . Why am I on that screen?" J. K. Rowling was aghast.

I froze. That was not what I was expecting.

Did I just lose my job?

I quickly removed the offending tweet and stammered out a frantic fan-girl reply, "Our sincere apologies . . . the mischief has been managed (aka quickly deleted)!"

Within an hour, I had generated more digital engagement than OIA had seen in the history of the organization.

It wasn't quite our target audience, nor did it have anything to do with the outdoor industry—but it taught me a valuable lesson as a social media professional: the internet moves fast and without mercy.

Despite J. K.'s quick acceptance of our swift apology, rabid Harry Potter fans flooded our mentions for weeks. The wrath of random tweeters haunted me, and I struggled not to take it personally. My boss was unfazed by the colossal calamity, but I had experienced what would soon become the standard experience for viral moments: content becoming bigger than itself and morphing into nothing but the distracting ping of phone notifications.

Eventually, the J. K. Rowling Twitter story would simply become a fun party trick to bring out among fellow social media marketers, but what happened was an indicator of how far and fast social media could reach. If we could piss off a tidal wave of wizarding fans, could we also use these tools for good to activate our own community?

The idea of advocacy is as old as humankind itself. Merriam-Webster defines advocacy as "the act or process of supporting a cause or proposal." Humans have always done this, enacting advocacy through literature, government processes, direct action protest, community organizing, resource sharing, and more. What if we translated those tactics to platforms like Instagram, Twitter, Facebook, email, and blogs?

Explaining the value of social media was an uphill battle in those days. Perceived by industry elders as frivolous millennial fodder, social media was largely dismissed as a critical element of communication plans. Those who understood its potential knew that it was the future of information sharing and connecting with communities. My boss at the time, Jenn Brunson, was one of those people. Despite skepticism from upper leadership, she persisted in supporting my work. And our program's growing success with every digital campaign proved that social media was worthwhile.

In digital, or social media advocacy, we take the elements of a grassroots movement and translate them to fit the platforms of the internet. Words and resources are typed up into PDFs, infographics, and tweet-sized captions. Stakeholders convene over Zoom, eliminating the restrictions of geographical space. Protests take the shape of coordinated Twitter moments tagging politicians with demands. When plucked from the physical world and converted into cyberspace, advocacy is made exponentially more accessible.

Social media always came easily to me. Some folks are born with a knack for music or art or athletics—I was made to tell stories with words and images. What is social media if not a new form of storytelling? Every caption, tweet, newsletter blurb, and live stream is an opportunity to craft a narrative and build a connection.

Anyone with a smartphone or internet connection can access social media and build their own accounts—aka personal platforms—which means everyone online has an opportunity to have a voice.

Advocacy is when we use our voices for the causes that matter to us.

Digital advocacy is when we take our voices online and adapt them to the platforms we use to be heard.

THE TWO TYPES OF ADVOCACY

Distilled down to the very basics, there are two types of "advocacy/activism" we're given the invitation to partake in. The best movements incorporate both, but ultimately, you ought to pick one side of the advocacy spectrum to prioritize for skills building. Before we begin mucking through the mess of how to balance being an advocate with being a human, it's critical to have a baseline under-standing of the two avenues for advocacy.

1. Digital advocacy on an infinite scale, first.

I will forever be a supporter of the power of digital advo-cacy. The undeniable top perk of digital advocacy is its exponential ability to scale and connect. Advocacy through platforms like Instagram and Twitter has con-nected me to leaders and community stakeholders in places I would have never had access to otherwise. I've hosted digital advocacy workshops with attendees span-ning from Alabama to Africa. It's because of the digital space that we were able to reach more than 12 million vot-ers during the 2018 #VoteTheOutdoors campaign.

The internet is a huge place and a powerful tool. We're able to distribute information in an instant. Spreading the word takes just a few fingertip taps. Need to fundraise? There's an app for that.

That said, this is the style of advocacy folks have been leaning into loudest during this age of handheld-every-thing, and to be blunt: it's not working. At least not like we thought it would. Not how we intended. Digital advo-cacy cannot be the destination—our actions online must always be leading us to action in the real world.

Digital advocacy also has a dark and rather conta-gious underbelly: the comment section and performative posting. Did that make your stomach clench a little bit?

Mine too. We've (that's all of us), begun to believe that activism looks like hopping online and engaging with the most insufferable bits of humanity.

You know those comment sections I'm talking about. It's that Facebook thread from the unhinged former high school cheerleader who drank the QAnon Kool-Aid, and you've anointed yourself deliverer of The Truths That Should Change Someone's Mind Even Though They Are Clearly Uninterested in Logic at This Time. It's the person on Twitter who continually baits you about politics, and you find yourself hashing out witty responses to their antagonization for umpteen hours. You know. We all know. We've all been there.

The unfortunate truth is: time spent in the comment section is often valuable time wasted.

This is not to say that all efforts to persuade folks toward different perspectives is useless—in fact, according to a Pew Research Center survey from 2020, "23% of adult social media users in the United States—and 17% of adults overall—say they have changed their views about a political or social issue because of something they saw on social media in the past year."

To give ourselves the full view of these numbers, though, we must reconcile with the other side of this data: "a majority of U.S. social media users (76%) say they have not changed their views on a political or social issue because of something they saw on social media in the past year."

We can see that roughly three-fourths of our efforts to change someone's mind online are going to be fruitless. That's a clear and abundant majority.

Complementary reporting from Pew found "64% of Americans say social media have a mostly negative effect on the way things are going in the U.S. today" (source: https://www.pewresearch.org/fact-tank/2020/10/15/64-of-americans-say-social-media-have-a-mostly-negative-effect-on-the-way-things-are-going-in-the-u-s-today/).

We must reckon with the diametrically opposed sides of this digital advocacy impact spectrum. The internet age gives us access to information distribution, collaboration across state and national borders, and communities of shared values. It also insists upon toxic communication cycles, proliferation of false information, and harmful community behavior.

There is a balance, a healthy line to be found when it comes to engaging with digital advocacy. Let's explore the other side of the advocacy engagement spectrum.

2. Advocacy as neighbors in real life, first.

Have you ever heard of a free community fridge? Salt Lake Mutual Aid started one during the pandemic with a simple concept: there's a community-supported fridge and pantry; come give or take food!

Volunteers set up two community fridges in Salt Lake City, which are public refrigerators located in volunteers' backyards and used to share food and ideas at the neighborhood level while reducing food insecurity and waste. They also aim to promote equal access to healthy foods.

When we prioritize advocacy through the lens of being good neighbors, we're entering a space of community care, generosity, celebration, witnessing, support, and reciprocity. I've found that my proverbial cup is refilled quicker when I engage with in-person advocacy. The stoke and commitment of others create an energy better suited to sustain us than the quick dopamine hits of social media.

Theoretically, hands-on advocacy is the way to go when it comes to direct impact, personal fulfillment, and community sustainability—but the physical world remains bound by space and time. The greatest strength of this form of advocacy, its in-person requirement, is also its ultimate limitation. In the real world, even our loudest voices only carry so far.

Perhaps our advocacy is the most powerful when we take the experiences and relationships from the physical and translate them to suit the reach of the digital. Social media platforms, the internet, emails—that's just how humans roll these days, and it's all a necessary part of existing in the advocacy space. But the digital shouldn't be our only destination just because our fingers have been wired to reflexively reach for it.

We can choose the spirit of advocacy as neighbors first, and still show up in the digital space. Recognizing that they are different, unique forms of advocacy is a key part of keeping ourselves accountable to upholding our values and practicing sustainable activism.

OUTDOORISTS

On a spring day in 2015, our team at OIA was brainstorming newsletter titles based on outdoorsy identities when my colleague Gareth Martins casually offered, "What about the word 'outdoorist'?"

We all lit up.

"Gareth, that's genius. More than a newsletter, let's make that the new term for how we identify ourselves in the outdoor industry!"

I declared that we would make "outdoorist" a commonplace part of the outdoor lexicon. Beyond a quippy email title, the term gave birth to a new identifier for our community.

More than eight years later, the word appears often in outdoor media. Folks include it as a personal identifier on their social media profiles. There is even a namesake non-profit now, the Outdoorist Oath, founded by some of the industry's top community leaders.

But back to the advocacy of it all.

As the digital mouthpiece for OIA, I became intimate with each of our department's work, especially the Government Affairs team. I learned that my tug toward a responsibility to caretake the outdoors could be channeled into impactful actions like calling lawmakers, lobbying on Capitol Hill, organizing petitions, and influencing policy as it's being made. If I was looking for a sign, that was it.

I decided to focus Outdoor Industry Association's digital identity primarily on our role convening policy and advocacy. Our platforms became speaker boxes booming updates about impending legislation, behind-the-scenes insights from our team in D.C., and perhaps most importantly, quick, accessible, easily digestible actions folks could take right from their phones.

Though our industry loves to tout the idea of wild disconnection and backcountry bliss, the truth is we're all on our smartphones constantly. As I was tasked with generating engagement around the organization's work on issues like permanent funding for the Land and Water Conservation Fund (LWCF) and protecting Bears Ears in Utah, I realized the key to unlocking advocacy action was handheld.

In the early days of social media, content was fun. Platforms like Instagram and Facebook were spaces we entered with the sole expectation of being entertained. It was the start of scrolling. Observing the amount of time my coworkers, friends, and family spent thumbing their way through the internet, I realized this new ease of access to digital platforms could be utilized for advocacy action.

If I had time to swipe through forty-seven blurry photos on Facebook from an acquaintance's summer road trip, I certainly had time to add my signature to an online petition. Anyone with the ability to slap an overly saturated filter on a pixelated snapshot of brunch was capable of navigating the basic skills of filling out a sign-on letter.

The key was meeting outdoorists where they were, rather than whistling at folks to get on our advocacy level. I paid attention to how we interacted with each other online, what types of posts got the most clicks and comments, and who was consistently getting a lot of engagement. The community leaned most toward storytelling, and the entire concept of the outdoors is ripe with stories.

On platforms that relied on photos and words, the visual of landscapes became a compelling backbone for outdoor advocacy.

While viewing images of place through palm-sized screens can never compare to seeing the real thing, being able to build connection to place without having to physically take someone there was a game changer.

I intended to use outdoor storytelling as the bait for attracting a sea of advocates.

THE OIA ROADSHOW

It took less than one year's worth of commuting in rush-hour traffic from Denver to OIA's office in Boulder every day to utterly fry my nerves and exhaust my spirit. There's nothing that can be said in defense of the daily commute—it's a slog and a soul sucker. I was driven to tears every single morning on my way to work, jerking my car back and forth in a chaotic vehicular mess.

I decided I couldn't do it anymore.

My heart felt homesick for life on the road, and I wanted to spend my time being present, experiencing the outdoor industry. Representing a multibillion-dollar community from beneath fluorescent cubicle lighting simply wouldn't do. I wanted to be among it. So I made my first big pitch deck.

The pitch: I was leaving Colorado to travel full-time, but I wanted to keep my job with OIA—and wanted the organization to support my travels as part of a major cross-country marketing and community engagement campaign.

The deck: Using PowerPoint, I created a twenty-page proposal laying out my plan to create and execute a field marketing trip aimed at strengthening OIA's membership. Yes, this was a good move for me—but it was also a winning move for the organization. This was taking field marketing to the next level for a scrappy little nonprofit.

I called it the OIA Roadshow.

In October 2015, I requested a meeting with my boss, Jenn Brunson, our marketing director, Jen Pringle, and our managing editor, Deb Williams. I was terrified, ribcage fluttering into my throat, but I launched into branded slides recounting my numerous successes at the Outdoor Industry Association. I reminded the women of the value I brought to the organization and outlined an extensive plan for social strategy in 2016, including a monthly calendar with daily tasks and deliverables.

Pausing on a slide with photos of myself and my coworkers in full OIA team spirit, cheering each other in matching T-shirts at an industry event, I dropped my plot twist:

"I'm leaving Colorado. I think you all know how unhappy the commute has been making me, and I've realized I really just want to get back to traveling full-time on the road."

Everyone's faces dropped.

"But I want to keep my job, and continue working for OIA from the road. And here's my plan for how I'm going to do it." I trembled as I clicked to the next slide.

The air was tense when I introduced the concept of the OIA Roadshow, but the skepticism quickly faded. My plan promised member visits, public lands tours, diverse content creation, and hosting regional events all aimed to boost our relationship with the outdoor industry community.

How would I join team meetings? How available would I be for quick questions on a daily basis? Would I still be able to attend all OIA events and trade shows? What if this doesn't work out? These were all questions I answered in my FAQ section in anticipation of pushback for this then-radical proposal for full-time remote work.

Since the COVID-19 pandemic began in 2020, remote work has become commonplace. But in 2015, the idea of letting a staffer transition to full-time travel while still maintaining their work responsibilities understandably left my employers with some skepticism.

Still, they said yes.

I left Denver in November and began the first leg of life back on the road. For the first few months, I focused on a beta version of the roadshow on the (mostly) East Coast. I spent four to six weeks each in three different cities, living in short-term rental housing. There was the musky apartment in Asheville, the carpeted manufactured house in Chatta-nooga, and the cute casita by the zoo in Albuquerque.

My one-woman roadshow visited outdoor retailers, toured gear manufacturing facilities, hosted sustainability-themed happy hours, and documented all of it on social media. I wrote trip reports for our website and sent our members all the images and videos taken while visiting for their own social channels. For the businesses, especially smaller operations, these touch points—along with the free marketing materials—provided a level of member-ship value previously unrealized by OIA.

Fueled by the success of the first season of the OIA show, I created a second pitch deck proposing a West Coast OIA Roadshow beginning in June 2016. This time, I wanted to live properly on the road, in a van—and I wanted my employer to pay for it.

Here's an excerpt from the pitch deck I emailed out to members I wanted to connect with and potential road-show sponsors:

Becoming an Advocate
for the Outdoor Industry

Outdoor Industry Association is based in Boulder, CO— but our 1,200 members are spread across the USA. As the leaders of this $646 billion outdoor industry, it's our duty to tell the stories of the people, places, and innovation that make up the outdoorist community. We're rolling up our sleeves and diving right in.

In January 2016, we sent our community and social media coordinator, Katie Boué, out on the road to visit members, host networking happy hours, and document outdoor industry culture across the country. The beta roadshow has been a success—so in June 2016, we're taking things to the next level with a West Coast tour. Over two dozen cities, five happy hours, and 6,000 miles of road travel.

> *"When OIA speaks, everyone in the*
> *outdoor industry listens."*
> *—Angie Houck, Darby Communications*

Why Should You Get Involved?

We're embracing the outdoorist lifestyle, and sending Katie out on the road #vanlife style. OIA believes in the power of collaboration, and we're inviting our members to hop in the co-pilot seat as we bring the outdoor industry to life through storytelling, events, content creation, and more.

> *Our roadshow will connect with everyone from CEOs to consumers, providing a strong platform for your brand to gain recognition and forge new relationships in the industry. Your brand will be highly visible to every retailer, manufacturer, and associate we visit.*

Needing a vehicle in which to travel, I sent individualized versions of the deck to over a dozen van companies with the promise that the OIA Roadshow van would be "the first thing industry leaders see as we pull into their headquarters, a roving landmark for outdoor community members who see us on the road, and a fixture at events." I'd add the van builder's logo to the title page of the deck and tweak the copy to reflect each recipient's brand. I wasn't just a dirtbag with a dream—this was business.

After a smattering of rejections and lukewarm interest, we secured a van. For the next four months, I would live full-time out of a carpeted Nissan NV 200. We slapped cheap branded roadshow magnets on the exterior, my then-partner drove me up to Mount Charles outside of Vegas for my first night of camping, and then I drove off alone.

It was one of the best summers of my life.

With everything I needed to live packed in a small van, I was a roving one-woman marketing department.

I visited dozens of outdoor brands that season, from mom-and-pop gear shops to international brands with security at the front door. I slept in the van nearly every night, finding whatever dispersed or developed camping on public lands would put me closest to the next day's meeting location. I slept out in nature, then would wake up crumpled in my sleeping bag and brush my teeth amid the morning dew. If I was lucky, I'd be able to avoid rush-hour traffic on backroads, but sometimes it was inevitable (read: literally anywhere in California).

My coworker Nikki Hodgson joined for multiple stretches of the trip during which we caravanned together with my dinky white van and her beefy gray Toyota Tacoma. She worked for the sustainability department, and we themed our happy-hour gatherings to reflect her work bringing the outdoor industry together on collaborative solutions and sustainable business standards.

Between her intimate knowledge of supply chains, chemicals management, animal welfare, and environmental policy, and my ability to curate a gathering and capture the accompanying stories that spilled from these spaces, we were an outdoor industry dream team.

At a sustainability happy hour event in Seattle, we packed out our reserved space at the REI headquarters. We traded business cards for free beer and raised funding to get youth outdoors through our Parks 4 Kids initiative. The crowd consisted of brand executives, local business owners, sustainability managers, outdoor media, and grassroots advocates—all of whom learned about this physical gathering through digital media.

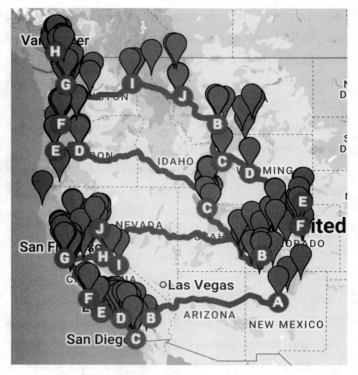

Map overlay of OIA members (pins) and proposed OIA Roadshow route (lines). Courtesy of the author.

That summer, I led my first traditional climbing route in Oregon with Nikki on belay; hung out with taxidermy fish while talking conservation at the Simms headquarters in Bozeman; caught a bird trapped in a Hood River café with my hands while taking a remote meeting; met hundreds of outdoor colleagues at networking happy hours I organized; learned about how supply chains and sustainability programs work at the Patagonia office in Ventura; and doing it all while finding somewhere to park and sleep each night.

One weekend between busy meeting schedules, Nikki and I decided to tackle six different sports in six different places between Portland and Seattle. We paddle boarded a river, climbed a few pitches, hiked to a summit, went for a trail run, and even rented surfboards for an afternoon that can only honestly be described as a pummeling by the sea. When the weekend ended, I stepped into our next meeting more fired up than ever on advocating for the outdoors.

My brain, body, and soul were spinning in a soup of purpose and passion—this is what I was meant to do. This is the kind of community and personal experience we're *all* meant to be living.

My personal writing started shifting from trip reporting to more passionate calls to action like: "It's up to us to start taking action + raising our voices to protect our outdoor spaces, and preserve them for the next generation of outdoorists. If you like to play hard, you ought to be advocating hard too" (source: My Instagram, August 29, 2016).

YOSEMITE + #ELPREZATELCAP

A few weeks into the OIA Roadshow summer tour, I got a hectic phone call from our headquarters. Nikki and I were sitting at the Crossroads Cafe in Joshua Tree eating a greasy breakfast when the ring came from our marketing director.

"President Obama is going to be speaking about the outdoor recreation economy at Yosemite in two days—can you be there in time?"

We stopped chewing.

"We'll be there!"

As we shoveled the remaining forkfuls of hash browns into our mouths, Nikki and I rescheduled our upcoming meetings and rerouted plans toward Yosemite National Park. We caravanned to the last hotel room left in Mariposa near the park, and as we passed through rural farmland my phone began to light up with a series of texts and emails fired rapidly enough to make me pull over to investigate.

"The White House just called and wants us to do a social media takeover, with rock climbers."

"Can you organize photos and captions from professional climbers like Alex Honnold and Tommy Caldwell?"

"We need to have this ready by tomorrow."

For an ambitious advocate hell-bent on the notion that social media could save us all, and with the promise of President Obama's blessings, my answer could only be, "YES. I'm on it."

With the sense of urgency and limitless stoke that would soon burn me out, I launched into go-mode. From a starchy hotel bed, I orchestrated a spontaneous event campaign called #ElPrezatElCap. (Yes, the hashtag makes me cringe now too.) We coordinated an entire day's takeover featuring professional climbers on the White House Instagram channel, and I managed to slip myself into the fold under the guise of representing the everyday outdoorist.

My caption read:

Hey everyone, @KatieBoue here! I'm not quite a professional athlete, but here's what I am: an American whose passion for public lands was inspired by climbing. I started climbing in the South, and dreamt about witnessing the

beauty of Yosemite National Park for years. When I finally
stood at the base of El Capitan for the first time, the magnificence of nature prompted me to dedicate my career to
getting Americans outdoors.

President Obama has protected more than 265 million
acres of land and water in our country—without his work to
support conservation and recreation, I wouldn't be able to
play in some of the places I love most.

Whether you're a pro athlete, or just an everyday
climber like me, the President's efforts to protect public
lands and make them accessible to all Americans is something to celebrate. I am truly honored to see #ElPrezAtElCap
this weekend.

The post gained 14,218 likes, and my phone lit up with
outdoor industry colleagues, personal connections, and
complete strangers all asking how they could get involved
with this outdoor advocacy thing. It turned out, under the
right conditions, even a mediocre climber could be influential over politics.

When social media democratized access to a public
platform, we all became influencers. The hierarchy of being
heard was radically disrupted, and suddenly anyone—
including a random young climber from Miami—could
become a public figure.

With a loud presence, knack for copywriting, decent
photography skills, heaps of privilege, and the type of

energy levels only a twenty-something could muster, I had as good a shot as anyone at being influential.

So do you.

As I watched President Barack Obama recite outdoor recreation economy statistics from meadows beneath Yosemite Falls, I felt the enormity of how impactful outdoor advocacy could be.

Clifftop security sharpshooters proved the enormity of this event. The idea of protecting the outdoors, of the huge impact the outdoor sector has, it wasn't just a niche idea anymore—it was coming straight from the president's mouth, echoing through Yosemite Valley.

On the shuttle bus back, I sat sandwiched between professional climbers and community leaders, pondering how I had ended up on this invite list. I realized that why I ended up here wasn't the point. What mattered is what I planned to do with the opportunity this had sparked. People were paying attention to the idea of outdoor advocacy, and I could be part of making this moment into a movement. The kindling was lit, but it needed to be flamed.

In July 2016, we hit 10,000 Instagram images tagged #outdoorist.

In January 2023, the hashtag hit over 51,000 posts.

Maybe you'll go tag a new post today.

END OF
THE ROADSHOW

That August, I attended the Outdoor Retailer trade show in Salt Lake City. I hosted a social media advocacy workshop and a luncheon to gauge interest in future advocacy projects. Both were standing room only.

Advocacy became a buzzword floating around every hallway and conference ballroom at the show. Folks wanted to get involved, and people were clamoring to embody this idea that the outdoor industry had a responsibility to act as advocates for the places we play—and frankly, the places that outdoor gear brands profit directly from. If you're selling hiking boots, shouldn't you be invested in protecting the trails?

During this moment of piqued curiosity around getting involved with the budding digital advocacy movement, social media was the key to unlock it all. Outdoorists across the country were able to connect, build relationships, collaborate on issues, and exchange ideas across state lines—all from handheld devices. We discovered that while our local landscapes may have different terrain, the issues of environmentalism, community, and industry could apply across topographies.

Most importantly, we found that the work of advocacy felt really, really good in the company of others who share the same values and stoke.

As my time on the road began to sunset, my visibility online ignited.

After introducing myself hundreds of times, shaking uncountable sets of hands, hosting five happy hours, and traveling to communities across thousands of miles in nine months, I had made myself a recognizable name across the outdoor industry. I had also picked up a considerable amount of Instagram followers as folks wanted to continue watching my summertime joy ride across the West.

Each stop on the roadshow presented new opportunities to build connections, create content, and amplify the word about what I was doing to spread the good tidings of bringing the industry together to do good. I recorded video interviews in company conference rooms, answered questions for journalists over GoToMeeting while sitting in hot parking lots, posed for photos with mixer attendees, and sat with many teams to discuss our shared visions of how the outdoor industry could do better and become stronger, together.

By the time I emptied the rental van of my belongings and drove it to a dusty parking lot in Las Vegas for the last time, I had a contact list longer than a CVS receipt. My time on the road shined a light on what I was capable of accomplishing as a one-woman show. I had also exhausted my patience with the corporate working environment, so I quit my job and vowed to freelance on my own terms.

When they brought in someone new to replace me, she only lasted a few months. OIA then hired me back as a contractor, for twice my previous rate.

At the time, this felt like a compliment—but ultimately, I realized this reliance on any one individual's participation could be a key flaw for an advocacy program. Not only does it complicate matters of ego for the person, it also muddies the water for the bigger picture of the organization or movement.

The reliance I had built upon blending my human being personality with the brand I was representing was ultimately exhausting, and I now believe that strong work–life boundaries for anyone working in marketing or communications are key to a sustainable career. When I became my work, I became burnt out.

Don't do that.

You are always more than just the work you do, and that includes advocacy.

#VOTETHEOUTDOORS

As then-executive director of Outdoor Industry Association Amy Roberts described it:

> After an election cycle where we have seen unprecedented attacks on our national monuments and a failure by Congress to reauthorize the Land and Water Conservation Fund—the most successful public lands funding program in our nation's history—the outdoor industry is ready to make our voices heard and help elect Republican and Democratic candidates who will stand up for our public lands and who recognize the power of the outdoor recreation economy.
>
> The 2018 midterms are our moment and we are asking for every outdoor-minded business and individual to help to amplify this message on their brand and personal social media channels on Oct. 15—and continue a social media storm through Election Day on Nov. 6.

So, that's what we all did.

Our team rallied the entire outdoor industry, from gear brands to individual content creators, under the nonpartisan cry to elect officials who kept protections of public lands as the top priority.

The campaign was multifaceted with strategic plans for every level of stakeholder, from CEOs to brand ambassadors. The Government Affairs team handled the substance of it all: vetting candidates, choosing balanced endorsements, combing through legislation and voter projections. Our team analyzed which election outcomes would best serve the outdoor industry and created a plan to put those people in office.

At this point, the plan got handed over to me—and that's where the creative fun started.

Armed with a Congressional Scorecard, Voter Guide, lists of key candidates, districts to target on social media, and connections to powerful outdoor voices who wanted to contribute, I drafted the rough vision for what would become Vote The Outdoors.

First, I gathered the assets. When you're developing a digital communications plan, you want to start with a bird's-eye view of everything you're working with. Outdoor Industry Association's robust government affairs and marketing teams had already created a thorough packet of assets ready to be distributed. My task was to take these politically technical, slightly jargon-y pieces of content and turn them into digestible bites suitable for the fast-paced nature of social media.

For me, this election was personal. I had flaked out on my civic duty as a voter in the 2016 election with fistfuls of weak excuses about traveling, absentee ballots, and registration deadlines—and nothing says "Yikes, you really should have voted" like watching reality TV star and woman abuser Donald Trump take office.

According to the US Election Assistance Commission, only 63 percent of the US voting-age population voted in the 2016 presidential election. Midterms draw notoriously smaller turnouts. In fact, the US Census Bureau found that just under 42 percent of voting-age citizens voted during the 2014 midterms. This was an opportunity to contribute to shifting those numbers and getting more voters to the polls. It was my responsibility (source: https://www.eac.gov/news/2017/06/29/newly-released-2016-election-administration-and-voting-survey-provides-snapshot).

We launched our efforts proactively—a rarity in an industry that is often driven by reactive advocacy news. The first major touch point of the campaign came at the summer trade show, where I passed out Vote The Outdoors stickers like candy. We hosted a Vote The Outdoors–themed photo booth at one of the big off-site parties. Our booth at the show included touch-screen devices where attendees could explore the congressional scorecard. Everyone at Outdoor Retailer knew about Vote The Outdoors by the end of the week.

The work ramped up as we crept closer to Election Day. Smartwool's president penned a blog post pleading outdoorists to do research using our tools. NEMO Equipment posted our Voter Guide to their brand website. Backpacker's Pantry designed a suite of Instagram graphics about voting. Chaco publicly asked their customers to please join them in voting for candidates that support public lands.

On National Public Lands Day, I used the holiday to remind folks that the best way to celebrate public lands is to vote. Just weeks before the election, The North Face hosted an educational workshop and happy hour highlighting our work. OIA's web team added a pledge form to our website where folks could leave a message telling us why they were going to vote the outdoors. More than 1,000 people had signed their names by October (sources: https://www.smartwool.com/blog/vote-the-outdoors.html; https://www.nemoequipment.com/blogs/journal/vote-the-outdoors; https://shop-eat-surf.com/2018/09/oia-on-the-road-vote-the-outdoors-workshop-networking-and-happy-hour/; https://www.instagram.com/p/BpNl7SNB0e7/).

I designed mediocre graphics and built a toolkit. The main message: "No matter your political stripes, the outdoors brings us together." We orchestrated industry-wide coordinated social media moments, flooding our community's social feeds with the good tidings of voting.

To engage the broader outdoor community, we turned to influencers. I tapped into my contacts asking them to record selfie-style videos answering the prompt, "I'm voting the outdoors because _____." We also reached out to local advocacy organizations like Latino Outdoors, Native Women's Wilderness, and Save the Boundary Waters. OIA's content manager, Deb, even managed to corner legendary filmmaker Jimmy Chin and ask him to record a sound-bite. We hired an editor to create a one-minute sizzle reel from all the footage we collected (source: https://www.instagram.com/p/Bpml-2UhBUM/).

Vote The Outdoors 2018 lives on as the ultimate case study for using every tool in your advocacy toolbox. Advocacy thrives on diversity of thought, tactic, access, invitation—and for this campaign, we hit it all. The political proficiency and insider knowledge of the Government Affairs team combined with the marketing team talents, from email and social media to managing web content created a powerhouse of communal action.

We understood that if there was one thing that brought Americans together, it was the outdoors. Positioning ourselves as a bipartisan political player allowed us to transcend political stripes, moving people to vote for candidates who would support our collective values about public lands, conservation, and climate change. The communications formula we developed would become a blueprint for activating political action through outdoor-focused social media content.

I reeled in outdoorists with a pithy social media post, or they were compelled to click on an email sent by one of my colleagues. Next, they'd land on our interactive web portal that provided tools to help folks learn who their representatives were, get educated on key outdoor voter issues, and learn about congressional voting records. The parting gift was my toolkit, with easy ways to share on social media sprinkled throughout the web pages. The call to action? Spread the word and Vote The Outdoors.

Our comprehensive communications plan was a success. We reached more than 12 million voters across the country. The majority of OIA-endorsed candidates were voted into office (source: https://sgbonline.com/by-the-numbers-oias-2018-annual-report/). In a follow-up report on the midterms, the authors cited the birth of what they referred to as the "outdoor voting bloc." They also concluded that in "most competitive states and districts across the region, candidates had to be pro-public lands to win" (source: http://westernpriorities.org/wp-content/uploads/2019/01/Winning-the-West_2018.pdf).

OUTDOOR
ADVOCACY PROJECT

I wanted to find the antidote to "callout culture" in the outdoors. Instead of policing my own community, I wanted to call people in. The only way you're going to change someone's mind or their behavior is if you have a relationship with them. I believed that social media could provide the tools necessary to facilitate a level relationship building that would lead to cultural shifts.

There were communication and understanding gaps between consumers and brands and lawmakers. We needed to be holding decision makers accountable, and telling them directly what we want as an outdoor community. I wanted to make it easier for folks to tell 'em.

I wanted to empower people to take action for the outdoors.

Outdoor Advocacy Project (OAP) was created with a vision of luring people into the web of outdoor advocacy as a lifestyle choice. I envisioned it as a wide-mouthed end of a funnel, or the first cobblestone on a long path ahead. OAP was never meant to be the be-all and end-all. It was built to serve as an invitation, a catalyst, a beginning, a convener. I wanted to create a platform for the community to gather, get informed, and then act. Our community is incredibly powerful—but few people in this era of TikTok attention spans are going to spend the energy to take action unless they feel fired up about it.

I spent weeks scribbling notes and ideas, sending long-winded voice notes to the soon-to-be organization's cofounder—and my best friend—Amelia Howe. We dreamt about what it would look like for the entire outdoor space to come together, to realize and utilize our collective force.

Two climate scientists, Savannah Adkins Croft and Chiara Forrester, became integral parts of OAP's launch early on. Savannah was a student at Utah State University in Logan, Utah, when she contacted me. I softly rejected her generous offer out of imposter-syndrome shame, but she persisted until I agreed to let her help by writing science content. She ended up organizing more than a dozen student scientists who developed content around forests, ecology, and drought. On the day of OAP's official launch, one of the student scientists, Chiara, woke up at three a.m. to fly to Salt Lake City from Boulder, just to be with us as we clinked glasses in celebration.

The first week of OAP's public launch, our Instagram page grew to more than 10,000 followers. This was during the phase when accounts could only use the now-retired "swipe up" feature once they hit the coveted 10K number, so we gained an instant advantage once we were able to directly link people to actionable web pages.

On December 10, 2019, we put OAP into the world. Our launch message read:

> Outdoor Advocacy Project is here. We believe the outdoor community has a deep responsibility to take care of and give back to the land—and each other.
>
> It's more of a commitment, really—to stewardship, sustainability, political action, community, public and tribal lands.
>
> The resources needed to empower the outdoor community on these topics are largely already in existence—built by non-profits, policy wonks, scientists, conservation organizations, and community leaders. We're a powerful collective.
>
> But there is also an undeniable gap between these crucial resources and the hands of the outdoorists who need them most. Outdoor Advocacy Project seeks to shrink that gap, collecting all the tools scattered around our industry and digesting them into accessible resources that prioritize

scientific data, inclusive perspectives, and actionable learning.

We believe in the power of the outdoor community— will you join us?

CLIMATE RALLY AT OUTDOOR RETAILER

A few days after Christmas in 2019, I got a text from my friend, ski mountaineer Caroline Gleich. She shared that she had applied for a permit to hold an event in front of the Colorado State Capitol with up to 3,000 people and asked if I wanted to help her organize a climate strike.

I thought she was out of her mind. I'd organized small workshops, hosted standing-room-only speaker events, put together happy hours in multiple states—but I wasn't qualified to orchestrate an entire protest. There were permits to apply for, we needed a plan to get the word out, and who exactly is in charge of all those chants? Is there a designated protest song leader, or do the chants just materialize naturally out of the crowd? Who would make sure the audio works? What if no one shows up?

When Caroline got approved for the permit, it felt like one of those moments when you're climbing a route, know you won't stick the next move, decide to just huck for it— and then find yourself still hanging on. It's a brief moment of "huh, that wasn't supposed to happen," immediately followed by the overwhelming but exciting dread that you've got to see this through now.

So we saw it through.

Caroline launched into gathering the details around event logistics, run-of-show, and safety marshal requirements. I drafted a communications strategy and toolkit to help get the word out. Caroline even emailed Jane Fonda inviting her to join us, but she was busy.

Outdoor organizations heard about the rally and wanted to get involved, including Wilderness Society, American Alpine Club, Protect Our Winters, and 350 Colorado. Media publications like GearJunkie and *5280* magazine wrote articles about the upcoming climate rally, saying:

> This week marks the sixth Outdoor Retailer trade show hosted in the Mile High City, but it's anything but business as usual at the Colorado Convention

Center. On Friday, while outdoor brands continue to showcase their newest wares to media folks and potential buyers inside the building with Denver's beloved Blue Bear, there will be excitement of a different sort outside: a climate rally.

Organized by professional mountaineer Caroline Gleich and Katie Boué, founder of the Outdoor Advocacy Project, the Climate Rally 2020 will take place on the final day of the trade show. And, that timing isn't a coincidence. (Source: https://www.5280.com/2020/01/a-climate-rally-is-about-to-descend-upon-outdoor-retailer/)

As we painted a long banner for the front of the march in Caroline's living room one snowy afternoon, we brainstormed ideas for chants and signs. She captured a long, bulleted list of slogans and clever plays-on-word; then we invited folks who had RSVP'd to the event page to add to the lineup. It included more than sixty classic and creative ideas like:

- A million snowflakes create an avalanche
- I'm with her. (point to a picture of earth)
- Climate change is a man-made problem with a feminist solution
- Heart and snow (heart with a snowflake inside)
- There's no business on a dead planet

- Planet over profit is a myth perpetuated by the fossil fuel industry.
- There is no Planet B.
- How we treat people is how we treat the planet.
- More women leaders = better climate policy.
- Clean air is a human right.
- Protect public lands.

I designed and ordered pins to distribute before and during the rally, with quotes like "The outdoor industry is responsible to act on climate" and "I Rally 4 Climate."

This campaign began online through event pages and save-the-date graphics, but it all started to feel real when the sign-making parties started. First, climate organization Protect Our Winters designated an entire section to rally sign making at their annual party at the McNichols Building in downtown Denver. I spent the whole night there, passing out craft supplies, helping kids paint their slogans, and using scrap cardboard to create extra signs for rally goers.

It felt real. This wasn't just a digital marketing campaign, it was bringing outdoorists together in real life to take collective action and witness each other. We were all advocates gathered around the same cause, sharing a physical space, covered in paint splatters.

Patagonia hosted a sign-making event the next day at their Denver store, and Outdoor Industry Association lent space inside their education zone on the Outdoor Retailer show floor for rally prep. We scavenged cardboard from every trade show nook, cranny, and recycling bin we could find. Local artist friends donated spare paint, brushes, and markers. At every sign-making space, the energy between advocates was electric. Everyone at the trade show was buzzing with anticipation of the rally.

Alongside the stoke, I was nervous. Disrupting the trade show I had spent the last four years working for felt like quite a bold move for a people-pleasing, nonconfrontational person like me. I was worried I might ruin relationships, or worse, that I would disappoint and disrespect the outdoor industry leadership I had formerly worked under. Turns out, taking a stand through screens was a lot less visceral that doing it in person.

By the day of the rally, 1,600 people had RSVP'd to the Facebook event. Our small crew of Outdoor Advocacy Project volunteer staff gathered at a rooftop parking lot and began to unload our piles of cardboard signs. We fixed pins on our hats, jackets, lanyards. The round yellow buttons with block lettering said, "The Outdoor Industry Is Responsible to Act on Climate."

The elevator ride to the ground floor was the loudest silence I've ever heard. No one knew what was about to happen, but it felt big.

We gathered at the base of the iconic Big Blue Bear statue at the entrance of the Colorado Convention Center.

We spread extra signs around the base of the bear for attendees to take and welcomed folks who began to gather in a boisterous crowd. As the group size swelled, Caroline passed megaphones to our rally speakers who started teeing up the spirit of outcry. An Indigenous man who goes by Big Water bellowed through the speakers as the crowd formed a circle around him. A call-and-response chant amplified the volume of our rally from one boombox voice to a cacophony of many.

"Protect the sacred," he yelled.

"Protect the sacred," we responded.

"Protect the water!"

"Protect the water!"

"Fight for our rights!"

"Fight for our rights!"

"For our sons and daughters!"

"For our sons and daughters!"

Golden retrievers ambled about with little cardboard squares tied around their necks declaring "Pups 4 the Planet!" Children rode on their parents' shoulders to get a better view of the growing crowds. I saw CEOs from outdoor brands, old friends, young folks, students, industry veterans—all convening together, at this moment, on the same issue, because Caroline wanted to do something with her frustrations about climate change and decided to apply for a permit.

About thirty minutes before we were scheduled to begin our march, someone proposed entering the Outdoor Retailer trade show floor and making a protest lap through the booths and hallways, to make sure the entire trade show knew what was happening outside of the convention center walls. I froze, knowing this would be an irreversible threshold cross for my relationship with OR and OIA. The trade show didn't take well to disruptions or going-off-the-plan, and I knew leadership at both organizations was already uneasy about the rally.

Folks traded badges to give noncredentialed protesters access indoors, and Caroline led a small group through the glass doors, waving their signs proudly. Security briefly attempted to stop them, but with event badges around their necks, the protesters were allowed to enter the space and take to the escalators. I could briefly hear the muffled ringing of their chanting from outside before they disappeared into the bowels of the convention center.

Sometimes, I wish I had gone inside too. Often, to do the right thing, you have to do a hard thing. Instead, I stayed outside with the majority of our gathered crowd and continued the chanting and preparation for our march down the streets of downtown Denver. It was a vital role that someone needed to play, but I wish I hadn't taken up the task to avoid the discomfort of what was happening on the show floor.

Both organizations were going to be unhappy about it either way.

After a relatively uneventful but distinctly unwelcome parade through the show, the group reemerged—bigger this time. More folks had abandoned their booths, packed up their meetings, and joined the conga line of protestors through Outdoor Retailer. Now, it was time for us to hit the streets.

I joined the women-led frontline of the march as we held up the big blue banner Caroline and I had painted weeks before. I felt a little insecure standing at the front of the crowd, and in reflection, that was a good thing. This rally wasn't about me, wasn't here to stoke my ego, had absolutely nothing to do with me at all—and yet it felt so deeply personal and powerful.

That's what successful advocacy feels like to me now: prioritizing an issue that's so much bigger than us while still maintaining the personal, human connection to it all. I think often of the "I'm just gonna step aside here real quick" feelings that formed during the rally. Sometimes, stepping aside to allow others to step up is the best form of leadership we can offer. It felt small then, but I understand that it's a big way of being now.

Our wave of humans started down the block, and I was instantly met with the sight of a long line of police officers positioned in a row on motorcycles.

They're here to stop us or intimidate us, that means we're doing something important, but is this rally about to stop before we even begin our march? I thought to myself.

"Who is in charge here?" asked an officer.

Caroline and I looked at each other, wide-eyed. "We are."

"Great, we're here to make sure you all are protected from traffic and can cross the streets safely; we'll be at the intersections to stop cars for you."

We marched onward.

Watching, hearing, feeling the sensations of hundreds of people traveling together down Fourteenth Street toward Civic Center Park felt electrifying. Red and blue lights flashed from the police as we crossed busy downtown Denver thoroughfares, and our line of advocates stretched farther down the street than I could see. A spiritual force took me when the chants started in Spanish.

"¡Aquí estamos, y no nos vamos!"

"¡Aquí estamos, y no nos vamos!"

"¡Y si nos echan, nos regresamos!

"¡Y si nos echan, nos regresamos!"

Responding in my family's native language to a protest chant being sung at a rally that I helped facilitate felt like the ultimate embodiment of myself as a human being and myself as an outdoor advocate.

Hundreds of advocates spilled over the grounds of Civic Center Park, and I paused in the flow to witness the full breadth of this moment. As the river of humans carried onward, I read signs and faces that passed by.

"Climate change is REAL and it's happening on our Public Lands," read the sign held up by a redheaded woman in a lilac beanie. This was her first march, ever. Other signs read:

"Coloradans Care About Our Public Lands."
"Gardner: Support the CORE Act!" (He wouldn't.)
"Green business is good business."
"SKODEN—Save the Planet."
"Climate Change Burgled Your Gnar."

I paddled back into the crowd to cross the street one last time and ascend the steps toward the Colorado State Capitol Building. Along with our permit came a podium, and audio had been set up earlier that morning. The speakers formed the horseshoe backdrop on the building's main steps, and convened advocates filled in the rest of the driveway, walkways, and into the lawn. The chants continued.

One by one, our speaker lineup took the stage. We opened with a beautiful but brutal land acknowledgment by Eastern Shoshone and Northern Arapaho dancer Sarah Ortegon, which followed by readings from youth activists Haven Coleman (then thirteen years old) and Madhvi Chittoor (then nine). Outdoor leaders like Deenaalee Hodgdon, Clare Gallagher, Jeremy Jones, and of course Caroline, spoke too. I played emcee, and listened behind the "stage" as each person's perspective and pleas boomed through the makeshift audio setup.

My friend Forrest Parks rounded us out with an incredible spoken word piece they had written just the night before. I remember finding the printed copy of their essay left on the podium after the event, and thought to keep it forever, because its contents were so touching and such a visceral testimony to our community in this moment. It got lost in the shuffle, and Forrest's words will forever remain purely preserved in the present moment for those who were lucky enough to bear witness.

The Climate Rally 2020 reminded me what we are capable of. We, the people, the outdoorists, the community. I was reminded that we don't need permission to do what is right. We simply must do it, because we know what is just and we are capable of it.

We were televised by Telemundo, featured by local and national publications, live streamed across the country, and were heard by lawmakers sitting inside the Capitol Building that day. Colorado governor Jared Pollis signed an official proclamation declaring January 31, 2020, as Colorado Climate Action Day in recognition of our march.

After the rally, we launched a digital toolkit aimed at keeping the momentum of the rally alive through online action. It was catered to the everyday person, a target audience I often describe as "a first time REI member who just bought their first pair of Chacos." It provided various levels of action to accommodate everyone from someone who doesn't know what climate change means to an intermediate advocate wanting to take climate action to the next level. There were three main calls to action:

1. Learn who your representatives and lawmakers are, and contact them about climate issues.
2. Make personal lifestyle changes—not because individuals are responsible to fix climate change, but to live your values and fuel your advocacy stoke.
3. Get educated and level up your climate science understanding.

The toolkit included blog posts written and organized by Savannah, including Climate Science 101, How Science Says You Can Fight Climate Change, and How Climate Change Affects More Than the Weather. It also included actionable resources like "A Step-by-Step Guide to Voter Registration."

With our post-event wrap-up, we hit full circle. It started in real life with two friends painting a banner, gained visibility through digital amplification, culminated in a collective physical experience, then was given a place for eternity on the internet.

THE ADVOCACY RETENTION CYCLE

Conceptualized by Amelia Howe, 2020

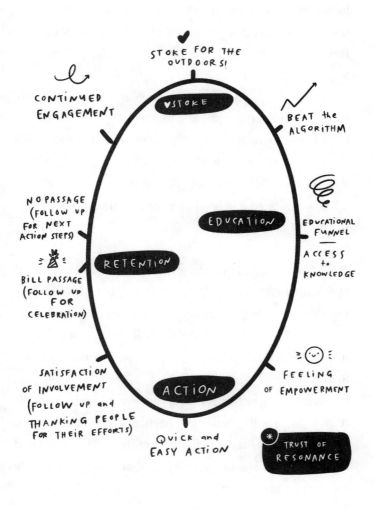

There's a fine artistry behind action alerts. They're those half-content, half-form web pages you click to when an organization says, "We need your voice, make yourself heard now!" Action alerts allow individuals to enter their information with a few keystrokes and automatically send it off to the appropriate decision maker's desk. The mastery lies not just in the ability to craft an easily navigable action alert, or in getting folks to add their name to it—the biggest hurdle to continued advocacy is getting people to come back and do it again (and again, and again).

How to retain advocates and maintain engagement on action alerts has been a commonplace conversation among outdoor communicators and organizers for years. It's such a critical concept that Outdoor Advocacy Project's cofounder Amelia Howe penned a concept called the Advocacy Retention Cycle.

At its core, the Advocacy Retention Cycle has four major steps: stoke, education, action, and retention.

It starts with **stoke for the outdoors**. "Stoke" can manifest a million different ways. It could look like one's passion for visiting—and therefore protecting—a particular place. A love for outdoor recreation or a desire to increase access for others. It might be activity based or community driven. It could even form through frustration around how a particular place is being managed.

Next, we need to **beat the algorithm**. The internet is full of noise, and the only way to capture someone's attention in the era of viral TikTok dances is to lean all the way in on the trends. This doesn't mean you need to start practicing choreography in the mirror (though it would probably help), but we must learn how to play the game if we want to win it. The prize in this scenario being someone's precious digital attention.

This is where the **education funnel** begins. Hook folks with a quippy, stylish graphic, then reel them in with digestible and pithy information. Providing **access to knowledge** democratizes the issue. By meeting people where they are—perusing around an aesthetic-focused, short-attention-span social media app—we're able to adapt advocacy to suit our target audience and alter the perceived barrier to entry.

Education fosters a **sense of empowerment**. Armed with the facts and solution pathways, advocates begin to take ownership over doing something about the issue at hand. With the necessary information and context in their toolbox combined with their personal story and voice, advocates realize that their voice is important—and that they are capable of taking action that may shift the needle on an issue they are passionate about.

One aspect that is not named as a core tenet of the retention cycle—but that instead permeates throughout each individual step of the cycle—is trust. It is key that when creating content that inspires action, we think critically about the sources of information we are providing and the language that we use. While the utilization of strong language creates an emotional response among advocates—and therefore assists in getting more clicks or impressions—misleading an audience can be detrimental to the retention piece of this process. As creators, organizers, and people on the internet, it is nonnegotiable that we do our due diligence and make sure that the messaging we run with is accurate and that we continually check in with the impacts of our efforts.

The climax appears next: action. The key is making it quick and easy. We're working with short attention spans, right? That means the process of engaging an advocate to take a specific action must be brief and simple. Minimal clicks, clear instructions, mobile-optimized platforms.

As we begin to descend from the climax, our key opportunity for retention is beginning. At this point in the cycle, folks need to feel satisfaction of involvement. This starts from the moment an advocate clicks "send" on an action alert—does the website offer a "thank you for taking action to protect public lands today" pop-up upon completion? Follow-up emails are crucial here too. A tactic we love to employ is engagement transparency and user-driven updates. During a 2020 campaign against a proposed oil and gas leasing sale on public land near Moab, we gave updates every time we hit a major milestone of signatures gathered. The original goal was 5,000 signatures.

When we quickly hit 5,000, we publicly announced raising the goal to 10,000.

Then 10,000 grew to 15,000.

Then 20,000.

Then 25,000.

Ultimately, we gathered more than 36,000 signatures. I attribute the exponential growth of the petition to the real-time, data-driven updates we provided throughout the campaign. Our community understood that their secondary actions of sharing the primary action item directly grew the moment and contributed to a bigger-than-just-one-signature impact.

The critical moment of **retention (or not)** comes when the campaign reaches success (or not). In the example of a piece of legislation, if it passes, follow up with celebration. If it doesn't pass the vote, follow up with continued action steps to alternatively keep the issue alive. Outdoor Advocacy Project's most viewed video was a simple animated graphic with the text "Breaking News: Great American Outdoors Act PASSED!"

In the caption, we wrote, "To our colleagues and co-conspirers who have worked tirelessly to support LWCF for the last decade: Thank you. To every outdoorist who e-mailed, called, wrote letters to your reps asking them to vote yes on the Great American Outdoors Act: thank you. Advocates, we have so much to fight for and so much more work ahead, but tonight, we celebrate."

That post was shared 3,641 times, saved by 875 people, and reached 46,870 accounts—all organically.

It feels good to feel good, and in a space where devastating losses are dealt often due to arcane politics, we need to relish every win we get. If we haphazardly skip from one action alert to the next without a moment to pause and look around, we lose sight of why we are signing these petitions. Victories aren't always on the menu for outdoor advocacy issues, but when we get one, having a moment of collective joy is a necessary step in preparing ourselves to charge ahead onto the next uphill battle. When we fill our cups during the victories, we are better able to endure the bitterness of inevitable losses.

INFLUENCE, PERFORMANCE, AND SUFFERING

Becoming self-employed shifted the way I worked. My efforts were dramatically pulled from largely behind-the-scenes to entirely displayed online for everyone to see and perceive. The heightened visibility brought a larger audience and a growing contingency of outdoorists eager to get involved with advocacy. It also came with an updated job description. Soon, the term "influencer" became part of my gig.

This felt like a blessing at first. People were paying attention to what I was talking about, and I could use social media storytelling to garner previously unheard volumes of action. A weekend camping trip to go hiking in the desert became a content package with videos, photos, and words all culminating in a clickable slide asking folks to show their gratitude to a place by emailing their senators about it. Folks couldn't add their names fast enough.

We all just want to be part of something good, and social media democratized the ways we can participate in community and advocacy. Hell, the link feature on Instagram stories alone revolutionized the way calls to action are accessed by the everyday human. Suddenly, we all had advocacy in the palms of our hands.

If this is what it meant to have influence, I was all in on being a cringey Instagram influencer.

As my reputation grew, so did the expectations strangers had of me based on my social media persona. The first generation of online influencers had no history to learn from or be guided by, so we just went for it. This would both serve and slam us in the end. Alongside my organic chronicling of my daily life, I started collaborating with gear brands on sponsored posts, going on free media trips, being cast as talent for outdoor advertising shoots. The perks were abundant, but little cracks began to appear—first in the form of anonymous hate mail or seething comments left by fake accounts. I continued onward.

During the summer of 2020, the call to use our platforms for advocacy was not just a casual rallying cry among the conservation crowd; it became an understood obligation we had as neighbors on planet earth. Outdoor advocacy is environmental justice is social justice. They are one and the same, all intersecting as part of a big web, and you cannot have one without the others.

That season we turned our attention from just protecting public lands for our own recreational pursuits to more expansive views of what it means to be an advocate. Advocacy was not limited to land and waters; it included our bodies and our communities. To post about protecting a landscape was no longer enough; we needed to consider the full picture of what it means to exist on this planet.

These radical shifts in thinking and heightened standards for advocating were necessary jolts of leveling up. It was the right thing to do, and we were capable of it.

The downfall started when ego got involved.

As you gain popularity on social media, you begin to blur the lines between civilian and celebrity. With a mere 60,000 followers on Instagram, I was hardly an A-List American icon, but I wasn't just a regular nobody either. I was receiving hundreds of messages each week from strangers. At first, I was determined to keep up communications. Being able to directly answer questions about political topics, helping people figure out how to register to vote, sending back encouraging notes to women saying they went on their first solo trip because of me—that felt good, and important.

The good feeling was great, but the sensation of importance would become a toxicity rampant among influential figures. In trying to do good in a time when everything was bad, we started trying too hard. I was considered an expert on outdoor and public lands issues, but from self-imposed obligation, I started feeling like I needed to speak as an expert on topics far beyond my capabilities.

The truth is: while I and everyone on the internet share a responsibility to speak up about injustices and use our platforms to help amplify issues, we are not qualified to act as spokespeople or experts on these topics. I am a white, able-bodied woman who has never truly feared for my safety around law enforcement. No one needs to hear my hot takes on police brutality. But in many instances, I, and so many of my colleagues, gave our takes anyways.

When someone sends you a direct message imploring you to "say something!" about an ongoing crisis, breaking news story, or community drama, you feel obligated to do so. Of course you care about this, so you should say something, right? After many years of taking the always-say-something route, I've learned to take a pause and ask an additional question: Am I qualified to say something about this, or should I just be amplifying someone else who is an actual expert?

Wanting to say something isn't the same as being the right person to say something.

To be fair to all of us who fumbled without grace through line between helpful amplification and just plainly being inappropriate, there was no precedent to follow. No formal systems of mentorship or check-in opportunities to say, "Whoa, that's weird dudes, let's do something different." We were all processing immense amounts of pain, grief, frustration, and isolation during that combined pandemic and social uprising. Some of us just made the mistake of documenting it all in real time on public platforms.

I was desperate to do good.

I did do good along the way, of course, but the road to hell is paved with good intentions, and eventually I began to feel my feet burning beneath me. We are told that social justice work is uncomfortable, which is true, but we aren't always offered the best tools to process that discomfort. When I should have been practicing journaling, therapy, and hard conversations with friends in real life, I instead dove deeper into oversharing on the internet and posting to prove my goodness. There's not a lot of shame left in my processing where it all went wrong, but I still grit my teeth and feel my cheeks flush when I think about posting videos of myself crying online. I was struggling to do good, to do it right, and I felt that if I showed how hard I was trying, it would prove my worthiness as an advocate.

Turns out, in all of this questing toward authenticity and vulnerability, I had started to fake it. The performance of suffering is an exhausting pursuit, and I wasn't feeling those this-is-my-purpose-on-this-glorious-earth vibes anymore. I just felt bad.

I believe that everything online is performative, and I mean that, but not always in a bad way. The content we produce is a form of art, the actions we take on social media are opportunities to stand behind our beliefs. Every word, photo, and post is an act of creative expression.

In other contexts, we readily accept the idea of performance as a good thing. A symphony orchestra creating a cacophony of sound with strings, percussion, and horns. An artist plein-air painting during a festival. That really cool dude who plays the spoons down on the street corner. It's human nature to perform; creativity is in our instinct.

That's why I say "everything we post on social media is a performance" without hesitation. But for most of my colleagues and friends, that statement is often received as an affront to the sanctity of our righteous causes. How dare we see our Instagram stories telling the tale of threatened public lands that need legislative protection as a means of expressing ourselves through creativity in the digital space? The same way we curate a cottagecore or VSCO girl aesthetic, we've been groomed into curating the display of the alleged pain we must show to prove our worthiness as activists. Yikes. If that's what advocacy is, no wonder we struggle to get folks to sign up to join us as advocates.

It is as if we're ashamed to have any joy in advocacy work. The world around us is suffering, and the internet has told us that in order to be good stewards of this imperiled planet and its people, we too must be suffering—and that suffering only counts if everybody knows about it.

If you ask me, there's already enough suffering on this planet. In fact, I find it disrespectful and audacious to add any more suffering to the mix. How dare we, as privileged, healthy, have-access-to-everything individuals perform additional suffering as if it is pushing us any closer to solutions for those who are actually suffering?

Honestly, how rude—to the earth, and to our neighbors.

The Great Salt Lake does not need me parading around on Instagram whining about how stressful and tedious my life is. Feeling paralyzed to inaction by the weight of the lake's direly low levels won't help either. And I am sorry to say, but completely bypassing the issue doesn't help the brine shrimp at all.

I am not advocating for us to ignore the suffering of the world, to carry on in merry ignorance, or to cash in on all our privilege and never look back. Those ingredients are not part of the recipe for healthy, successful, sustained advocacy.

Instead of absorbing and performing the suffering we see around us, I am inviting us to instead bear witness. We are so quick to make everything about ourselves—and if you've ever navigated a friend's tough time (breakup, bad news, disappointing results), you're familiar with how incredibly unhelpful it is to receive someone's pain and then immediately make it about ourselves.

Your suffering does not contribute to protecting the environment. Your woe-is-me attitude toward yourself doesn't help others around you.

The nuance here—because there's always nuance to be had—is that making outdoor and environmental issues personal is exactly what gets folks activated to take action. The key is making sure the appropriate pieces of pain and joy are being witnessed, and keeping the movement focused on the ultimate mission.

Bears Ears in Utah provides an excellent example of this.

In 2016, President Obama designed an executive proclamation declaring the establishment of Bears Ears National Monument (source: https://obamawhitehouse. archives.gov/the-press-office/2016/12/28/proclamation-establishment-bears-ears-national-monument). This action protected more than 1 million acres in the southern region of Utah held sacred by many Indigenous peoples, including the Navajo Nation, Ute Mountain Ute Tribe, Hopi Nation, Zuni Tribe, and others. The landscapes are home to rock art, ceremonial sites, thriving and precious ecosystems, abundant recreation, and a history dating back to time immemorial. Native communities had been rallying to protect this area for decades, and President Obama finally made it happen.

Despite the fact that the proclamation specified that current cattle grazing, timber, and private access rights would remain intact, some rural Utahns saw this public lands protection as an infringement on their rights by the federal government, and the topic was hotly contested within state lines and beyond. It was clear the pro-monuments side would have to rally together with unprecedented unity. Climbers, ranchers, scientists, historians, and adventurers of every activity joined together to strategically use our voices to protect Bears Ears—and the highest priority voice of all: the Indigenous communities.

On a planning call during the early stages of the Bears Ears campaigning, I was eavesdropping as key stakeholders from national recreation and conservation groups across the West discussed how to move forward in response to unfolding bad news. A few folks offered versions of statements but were interrupted by one man who said, "We shouldn't put out any statements or take any stances until the Intertribal Coalition does—and then we should follow their lead."

So, the outdoor industry largely did just that. We used our leadership skills to follow the lead of the land's original stewards, and the movement strengthened through collaboration. Outdoor communicators still knew their audiences and shared the compelling stories of why this political issue mattered to climbers in Indian Creek, paddlers on the San Juan River, and hikers who love slot canyons—we just did so as a method of funneling advocates toward widening their perspectives on why Bears Ears matters. Come for the climbing, stay to fight for Indigenous sovereignty.

Without the impassioned outcry and personal pain bled about the cultural, spiritual, and geological significance of Bears Ears, it wouldn't have been protected (and then unprotected, and then reprotected, but I digress). Across the country, Americans bore witness to the history of Bears Ears, the plight of Indigenous communities who have been violently kept from their homelands, the pain of generations.

The movement to protect Bears Ears remained focused on Indigenous storytelling and prioritized Tribal leadership.

When it didn't, the movement suffered. Lest we forget Patagonia's "the President stole your public lands" fumble when the brand blacked out their commerce site with the above message and instantly received a clap back from Indigenous communities and their allies reminding Patagonia that public lands were stolen from them to begin with.

But mostly, we kept steady on the mission to collectively bear witness to the abundant history and painful present facing those who call Bears Ears home, and instead of making it about ourselves, we used that energy to rally, educate, and act. There were, as there always are, many missteps—but mostly, we rose above our own egos for the greater good. That's called advocacy, folks.

Everyone on this earth—especially those who call ourselves outdoorists—has a responsibility to advocate for the protection of this planet. Our very existence relies on the earth around us, and we are indebted to it. It is our lifelong duty to leave it better than we found it.

We must also remain in service to our humanity at all times, no matter how dark the world around us seems. It is rude to create any additional suffering—is our planet's pain not enough for you? Perhaps imitation is the highest form of flattery, but who are you trying to impress? What is the purpose of usurping pain with more pain?

If we supposedly can't bear another ounce of pain on this planet, why are we so obsessed with performing our own suffering? As if our egotistical, infinitesimal problems were in the same league as the issues plaguing the environment around us. As if we could be so audacious as to compare our mortal maladies to the critical matters of the universe.

If influencing people's action is what I'm doing online, I want to influence people to do good and to feel good. So, it's a no for me on the suffering bit anymore. I've got better ways to spend my time—and so do you.

CRASH AND BURNOUT

Patagonia founder and outdoor industry environmental icon Yvon Chouinard once said, "The cure for depression is action. Every one of us has to step up and do what you can, according to what resources you have."

I deeply agree with that second sentence.

I have some small objections to the first.

At the end of summer in 2020, I was mentally unwell and spiraling toward a classic case of career burnout. By the time I was entrenched in the bowels of depression, I had blown past all red flags and opportunities to check myself before I wrecked myself, as they say. Despite daily crying fits, unexpected visceral flashbacks of old assault traumas, chronic fatigue and brain fog, obvious inauthenticity when showing up online, bouts of disassociation and growing suicidal ideations, I chugged on. It was an election year, and after my shameful no-vote in 2016, I felt responsible to make sure this election didn't put Trump back in office. I would later recognize that this savior complex helps no one.

I continued showing up for work, struggling through Zoom calls, eking out deliverables under the wire, designing half-assed graphics, lying down on the kitchen floor between each task and staring blankly into the floorboards, wishing I could cease to exist. And for what?

What I know now from an eighteen-month battle with depression, anxiety, and cyberbullying is that when you need a break, you *need* a break. If you do not give yourself permission to take that break, bad things will inevitably happen. I forced myself to keep suffering online, to continue feeling righteous enough about my platform (read: ego) to self-flagellate on a near-daily basis. I would have an anxiety attack about needing to record a video about an advocacy action alert, pull myself together long enough to post it, then go sit outside and cry under a pine tree for hours.

Along with a handful of other outdoor content creators, I had become the subject of vitriol and anonymous harassment from an online group that tracked my every tweet, story, post, and sometimes even whereabouts in real life. When I moved across town that summer, they doxed my new home address by screenshotting a video I posted of my new garden and combing Zillow until they found a listing that matched the shot. Any anonymous stranger was given my personal home location, just because some women on the internet didn't like my Instagram. I became obsessed with consuming the hate directed toward me. I began to believe that I deserved what was being said of me. This is what I got for being a public figure on social media. I started to believe I was a bad advocate—and a bad human. I checked the forums morning, noon, and night, stunning myself with pain every time I logged on.

If being online is making you miserable, get off.

Our obligations to perform digital activism will never outweigh our responsibility to exist first as decent, sustained human beings.

I had become indecent. Perhaps you know the feeling.

As advocates, we are meant to live complex, challenging, celebratory, curious, and creative lives. The issues we are dedicating ourselves to are hard, and often frustrating, but the sum of all the ups and downs of advocacy is meant to net out positively. We are not meant to carry all the burdens of the world; we are not meant to copycat the suffering we see around us. We are meant to heal—ourselves and the earth around us. Not all of it. Just the bits we can access.

That's okay. That's enough. We were never meant to fix it all, to be it all, to be able to comprehend and hold and witness it all. We're just big sacks of skin and sticks.

I started dropping the ball on projects that once brought me stoke, receding instead toward isolation. Editors' emails asking for pitches went unanswered, projected campaign finish dates postponed into oblivion, then I stopped responding to messages from friends and colleagues. Eventually even phone calls from my own parents felt like more than I had to give.

When we convince ourselves that our continued suffering is part of "doing good," that isn't the righteous cause speaking—it's ego. No one asked me, or you, to perform suffering as part of contributing our portion of shared stewardship. No one.

My therapist once described the scene to me in vivid and brutal detail: she said I had crucified myself with importance. In the name of "doing the work," I had erected a giant metaphorical cross, crawled up onto it, hammered myself to the wood, and was now just stuck baking in the blistering sun, wailing out in nonstop pain. The people down on the ground, whom I had crucified myself for, were absolutely unaware and could not care less about my miserable show. I was suffering on display for absolutely no reason. No one had asked for it. (To be fair, anonymous internet trolls *had* specifically asked me to suffer and show it, but as my therapist quickly reasoned: anonymous losers on the internet are literally no one.)

It was a slap in the face to hear that, but she was completely right.

When we fool ourselves into believing that we must suffer in order to be good advocates, we are choosing ego over purpose. I had begun to think this was what accountability looked like. The moment we choose suffering, we have lost our way on the path of advocacy.

I had wandered far from the path and found myself in spooky metaphorical woods, completely alone. No one had abandoned or exiled me—this was just a symptom of my unhealthy relationship to my responsibility as an advocate. Instead of seeing my isolation and distance from colleagues and loved ones as a point for course correction, I interpreted it as part of the punishment I had to endure in order to be worthy as an advocate.

Folks, that was incorrect. First, advocacy is a community event. Second, performing suffering and misery by yourself online isn't a viable personality and doesn't contribute to solutions, especially not in conservation or climate communities.

Humans are meant to exist in togetherness. We are meant to hold these heavy and complex issues of our planet with many hands. These burdens are too heavy for any one of us, so we distribute the weight amongst us all. Advocacy succeeds when we operate like a beehive, every member understanding their individual role, caring for the overall colony, and remaining focused on the mission—to make honey, or pressure senators into passing legislation.

My self-inflicted suffering wasn't just detrimental to me, it was harmful to the entire outdoor advocacy hive. The result of my pain was ego, and I have learned that especially on social media, ego is toxic to movements meant to be held in shared community.

In hindsight, I wish I had logged off Instagram earlier.

I finally did, though, in October 2021. I had begun the healing process, and felt capable of starting to show back up to the idea of advocacy with a renewed sense of curiosity, eager to examine and correct my missteps along the way. The first big step in this was to cancel all upcoming partnerships and take an extended break from my role as @katieboue, verified outdoor advocate influencer on Instagram.

I gave myself permission to remind myself what advocacy feels like out of tiny handheld squares. So, I just stopped posting one day. I wrapped up one remaining project, a reel about an in-person trail maintenance day I had attended with Tread Lightly and Visit Utah, which felt apropos, and then I was gone. I briefly considered making An Announcement about my departure from the 'gram, but instead opted to just slip away into silence.

Offline, I reimagined how I could spend my time and energy. I slept, a lot. Though the season quickly dwindled with the coming of winter, I spent more time in my gardens than ever. I started to reimagine my relationship to the outdoors and give myself permission to just be. When tragedies happened in the world around me—and it turns out they do, often—I processed the pain and galvanized toward local action, with my friends and family keeping me accountable, in real life. There was no pressure to perform, because it turns out, our digital performances are just a show—and shows are for spectating, not living.

Real life will always usurp the digital world.

And it should.

What a relief.

DEHUMANIZATION AND EXPECTATIONS

It's a strange thing, to no longer feel like yourself—but that's what happens when you sell your soul to Instagram. When your entire livelihood hinges upon the performance of your*self* for others, you eventually become more performance than yourself. It's not to say that one day my soul stuff was suddenly ripped from my physical body, or that there was some dramatic moment in time where I stepped across the threshold of no return to influencer hell. It happened over time, and I was always in there, barely, but there.

The careful tucking away of the self into the most private, hidden spaces is a symptom of the world around us providing constant feedback that the weirdest, rawest, most unusual bits of you are not suitable for public consumption (and that might ultimately be right, but I digress).

Those parts of me that are most myself became wea-
ponized by snarky trolls who assisted me in the twisted
mental process of conflating who I am with the job I have.
The way I speak, my Miami accent, my proclivity for the
word "bro" as a response to any occasion, the white eye-
liner I've worn daily for the last fifteen years, my queer
identity, my Cuban identity, the scrunched-up exaggerated
facial expressions I've perfected after years of responding
to *chisme*. These things that are mine, the little bits that
made me myself, suddenly weren't anymore. They were
fodder for declaring me a bad advocate.

I began the work of shrinking down my big personality to blend into the outdoor industry from the moment I
logged onto social media in pursuit of sponsor support for
my yearlong van trip in 2012. Perhaps it had started even
earlier, assimilating my loud, big personality into smaller
and smaller spaces, hoping to fit in with the cool guys at
the climbing gym.

Be palatable. Be easy to work with. Better yet, be a
sample size. Make yourself big to fit our campaign and
promote it, then make yourself small because you don't
deserve to take up too much space here.

My entire life has been a balance between too much
and not enough. I acknowledge the precarious place I sit
perched in identity. A Cuban-Venezuelan second-genera-
tion immigrant who is white. A queer woman in a long-term
committed relationship with a cis man. I was born cloaked

in privilege, after my parents carried the lifelong burden of breaking generational trauma and embodying the "American Dream" so I could have a better life than they had.

I am a contradiction and a mess and more than a little bit nuanced—and if there's anything I've learned from my decade of working in social media and digital community spaces, it's that the internet does not expand to fit the needs of nuance. It simply is not the space for it, necessary as it is for productive dialogue and societal progress. And yet, we spend our entire digital lives convinced that if we just add another caveat to this caption, reply to that comment for the third time to clarify, suddenly we will burst with the consideration, flexibility, and curiosity needed to achieve the proper nuance.

To be fair, the time it would take to seriously consider, absorb, digest, and react properly to someone's entire identity or big philosophical claims made online isn't exactly how I want to be spending my time on the 'gram. I am a very weird, complicated, not-for-everyone, bubbly manatee of a person, and there's no feasible way to communicate all of *this* through a small photo or video transmitted through handheld screens. I'd be a pretty small, uninteresting human if you could take in all of me through a patchwork of posts the size of your palm—and so would you. So why bother trying to cram entire people, places, problems, philosophies into these tiny squares? Where's the fun in that?

I'd rather let these squares be the patchwork that they are. Snippets and snapshots and quick reference guides to the too-big-for-a-fifteen-second-reel real-world expanses that are our lived experiences. I can't fit into a tweet the full scope of how impactful it was for me to be part of a relay run through Bears Ears National Monument across more than 250 miles and three days, so I'm going to stop trying. Instead, I'll share what I can, and let it be just that: moments in time, smatterings of words I felt inspired to write, quick notes in photos and videos to capture moments that I might like to be reminded of one day.

These aren't just selfish acts turning our entire feeds into nothing but personal photo dumps. This is the continued and honest work of storytelling and humanizing the advocacy work we believe in so deeply. We can do it more authentically this time because we're realizing that we don't have to try so hard to be the perfect advocate and solve [insert crisis here] every time we log online.

#RECREATERESPONSIBLY AND THE REDEFINITION OF SOCIAL MEDIA CAMPAIGN SUCCESS

In the midst of my spiritual catastrophe, an unexpected salvation brought me back into focus: the Recreate Responsibly Coalition. Born during the height of the pandemic, the Recreate Responsibly Coalition was a reactive effort by public lands management agencies and other outdoor sector stakeholders to create unified messaging during a time when directive advice and authoritative expertise were hard to find, let alone heed.

The coalition formed during the first days of the COVID-19 pandemic, when public lands simultaneously shuttered and became the only safe space many folks had to turn to. Navigating shelter-in-place mandates, rapidly evolving hygiene and safety recommendations, and conflicting advice on what we should all be doing—it was a swamp of information to muck through. Nearly everyone working in the public-facing outdoor space found ourselves with an instantaneous and severe need for community guidance and standards across state lines and land destinations.

Through countless meetings, and the best group edits of my career, we designed seven original tenets to capture the key messaging outdoorists needed to safely get outdoors during what we all hoped would be a brief pandemic.

The original Recreate Responsibly guidelines:

1. **Know Before You Go.** Check the status of the place you want to visit. If it is closed, don't go. If it's crowded, have a plan B.
2. **Plan Ahead.** Prepare for facilities to be closed, pack lunch, and bring essentials like hand sanitizer and a face covering.
3. **Stay Close to Home.** This is not the time to travel long distances to recreate. Most places are only open for day use.
4. **Practice Physical Distancing.** Adventure only with your immediate household. Be prepared to cover your nose and mouth and give others space. If you are sick, stay home.
5. **Play It Safe.** Slow down and choose lower-risk activities to reduce your risk of injury. Search and rescue operations and health-care resources are both strained.

6. **Leave No Trace:** Respect public lands and communities and take all your garbage with you.
7. **Build an Inclusive Outdoors:** Be an active part of making the outdoors safe and welcoming for all identities and abilities.

By noon on launch day, the #RecreateResponsibly social media campaign had reached more than 5 million people. In the next eight months, we'd hit over 4.5 billion (that's billion with a *b*) impressions.

During the first year of the Recreate Responsibly Coalition, we convened more than 1,300 partners across public lands agencies, nonprofits, outdoor brands, other federal governmental agencies, and more. One national coalition birthed multiple state chapters. We galvanized numerous working groups based on niche expertise, and as a result launched special Winter, Wildfire, and Off-Road guideline editions. Our standard guidelines were translated into nine different languages.

As the global pandemic evolved, so did our messaging needs. Staying close to home morphed into guidance about exploring locally. Advice about practicing physical distancing was replaced with calls for respect and building a more inclusive community.

The most recent iteration of the Recreate Responsibly guidelines:

1. **Know Before You Go.** Check the status of the place you want to visit for closures, fire restrictions, and weather.
2. **Plan and Prepare.** Reservations and permits may be required. Make sure you have the gear you need and a backup plan.
3. **Build an Inclusive Outdoors.** Be an active part of making the outdoors safe and welcoming for all identities and abilities.
4. **Respect Others.** There is space for everyone and countless outdoor activities. Be kind to all who use the outdoors and nature differently.
5. **Leave No Trace.** Respect the land, water, wildlife, and Native communities. Follow the seven Leave No Trace principles.
6. **Make It Better.** We all have a responsibility to sustain the places we love. Volunteer, donate, and advocate for the outdoors.

My greatest joy in my work with this movement is how utterly insignificant our own "brand presence" is in the success of the campaign. Recreate Responsibly generates billions of impressions and reaches millions of users across the United States (and in other countries too), but a truly minimal amount of that engagement comes from our own social channels. Our brand's identity isn't the point—the message of the movement is what matters here. We aren't the harbingers of the guidance because we aren't the sources of authority or community trust. We're just the engineers to build the speakers. The coalition members themselves are the voices transmitting through the microphone.

In a time of deep ego struggles, this work was a reminder of how powerful we can be when we are capable of separating ego and self from advocacy. I don't get constant credit for this work, nor do I want or deserve it. I'm simply a collaborative convener on Zoom calls and a talented social media marketer. I have skills in the communication space, and I use them.

Our small but nimble team has developed a well-oiled flow for communication drops. Whenever a new edition is released, we give it a thorough treatment that produces everything from suggested social posts to downloadable graphics. The goal is to make coalition-wide amplification as easy as possible in hopes of high adoption. All you have to do is click one link to our toolkit, and you're handed easily copy-and-pasted captions for various social media platforms, pointers on hashtags, and links to graphics and PDFs.

As the popularity of the outdoors continues to grow, so does our opportunity to rethink the responsibilities the outdoor community has to each other and to the natural spaces we occupy. So many of us are reimagining our relationship to the outdoors where millions of Americans found respite during the darkest days of the pandemic. The Recreate Responsibly Coalition sees this opportunity and understands the duty the outdoor sector has to provide leadership and help shift our culture to better reflect our values.

One day, a hotshot at Facebook emailed the coalition pitching an idea about creating a special edition of the Recreate Responsibly principles, designed specifically for creatives. Over many emails and a year's worth of organizing calls we launched the Create Responsibly initiative. The aim is to develop a community of creators who harness the power of their resources to amplify responsible content creation practices and a sense of shared stewardship.

The Create Responsibly guidelines:

Know Before You Go. Research and contact the location well in advance. Connect with land managers about permit requirements, weather, and local guidelines.

Plan and Prepare. The outdoors can be high risk with conditions that change quickly. Be ready with the right gear and be aware of drone "no fly zones."

Build an Inclusive Outdoors. Inspire others to see themselves in nature. Showcase diverse backgrounds, abilities, and meanings of the "outdoors."

Respect Others. Minimize impacts to other users. Avoid blocking trails, and give space to people recreating or working.

Leave No Trace. Respect the land, water, wildlife, and Native and local communities. Avoid sensitive habitats. Showcase responsible use.

Make It Better. Inspire others to respect and nurture our outdoor spaces. Depict responsible recreation in action—and behind the scenes.

We launched a pledge page and collected information from creators who were interested in joining the movement. Our philosophy is that creators are responsible to the outdoors. Sound familiar? The foundation of what we share out into the world should be built upon social and environmental responsibility and equity across platforms. As folks who are often creating on public lands, we directly benefit from those lands and are therefore responsible to care for them.

From the Create Responsibly working group to multiple state chapters, the Recreate Responsibly Coalition has evolved into a complex ecosystem with space for everyone who touches the outdoor world. Every day, new members join the coalition and working groups. As of December 2022, there were more than 88,000 posts on Instagram tagged #RecreateResponsibly.

As the tactics and community grow, the core mission remains the same: to positively influence outdoor behavior and expand ecosystem understanding through consistent and accessible messaging, platforms for collaboration, and elevating voices to ultimately inspire compassionate stewards of our communities, land, and water.

PERCEPTION, SCALE, AND IMPACT

After running multiple organic social media campaigns reaching millions—and even billions—of social media users, my standards for a successful campaign have risen considerably. That's the goal on social media: more, bigger, and instantly.

The viral promise of infinite reach coupled with the instant satisfaction and microscopic attention spans fostered by social media means we've lost touch with how meaningful every single individual affected by your advocacy is.

If I got twenty-five "likes" on an Instagram post, it would feel like the end of my career. According to the Instagram auditor I just plugged my username into, my average post attracts 3,200 likes. This means my expectation is that every time I post something on Instagram it will reach and impact thousands of people. Thousands.

Anyone who has met me in real life understands immediately that I am a weird, awkward person in social situations. I don't do crowds. My girlfriends and I have a running bit where we select a "party bush" upon arriving to a venue, because we all know I will inevitably need to excuse myself and go hide in a bush at some point when I become overwhelmed with socializing. So imagine me, in real life, doing something and *thousands* of people showing up to let me know they liked it.

That does not sound pleasant in any way.

So let's revisit the scenario above: If I got twenty-five "likes" on an Instagram post, it would feel like the end of my career. However, if I hosted a workshop or free garden share stand, and twenty-five people showed up—that'd be a clear success. Twenty-five is a veritable crowd!

Twenty-five individual human neighbors, in real life, sounds pleasant indeed.

I had become so accustomed to the grand scale of social media that I lost touch with the human scale advocacy is meant to operate on.

Is there a universe where we can have both robust in-person communities that focus on quality and local relationships, and infinitely viral social platforms that help us spread the good tidings of the real world? I think so.

After eight months of abandoning my main digital megaphone and solely focusing on more tangible, small-scale advocacy efforts, I believe both are necessary for optimized success as advocates. The real world coming together, action, and community provide the digital story-telling that helps us amplify our work beyond the reaches of our local movements.

It was difficult for me to adjust to the different scale of impact. Initially, it felt less important. How could I compare 10,000 people seeing my Instagram stories about killing your lawns to the eleven people who stopped by my little free farm stand to share garden veggies? The dismantling of notions created by ego is a blunt process. To offer yourself permission to shift your perspective into something new is a gift.

So I gifted myself the invitation to change my mind. Realizing that the quality of advocacy far supersedes the quantity of measurable reach was a powerful reckoning—and one that did not require me to shame myself or discount past perceptions of success as an advocate.

Reaching huge volumes of people on social media does matter, still. It always will, at least in our lifetime.

I'm still impressed by social media analytics detailing how a series of digital posts can reach millions of users in a limited time frame, no doubt. The more people who hear these stories, find inspiration in them, find applicable resources for their own pursuits, the farther advocacy spreads. But the ultimate point is never the digital.

The point is always in person. We do not strategize, create, and execute advocacy campaigns for the analytics report. The ultimate goal is resources delivered to someone's hands, a signature on paper protecting land, boots on the ground stewarding a place, humans gathered together.

When distilled back to its purest form, advocacy operates on a human-sized scale, not the scale of the World Wide Web. That's okay. In fact, that's excellent. That's the way it was always meant to be; we just lost our perspective when the Instagram dopamine hits started coming in rapid succession.

Lessening our expectations about scale of reach allows us to focus on quality of impact. While casting a net far and wide has strategic purpose within campaigns, what really keeps an advocate coming back for more is the depth to which they are affected.

I care less now about how many sets of eyeballs saw an educational graphic scroll across their feeds. I'm more keen to know how many people called a friend to talk about it. The number of times an event registration link was clicked is cool, but tell me more about how many sets of hands showed up to support the trail day. I love being able to count how many people opened an online toolkit, but I wish I could watch their minds churn while implementing those tools in their own pocket of the world, adapting resources to serve their unique community.

When we shrink our circles to better meet the capacity of our mortality, we're expanding our capabilities for deeper impact around us.

THE SWELL, UTAH

The biggest impact I had on the outdoor place I love most garnered the fewest eyeballs of any project I've ever worked on before—and that lack of virality was a measurement of its success.

Nearly ten years after the city cleanup day in Joe's Valley, my friend Adriana, who was there that day, was now serving as the county's director of tourism. She also helped found the Joe's Valley Climbing Festival, an annual celebration that brings together rural life and the climbing community for a cultural exchange unlike any other. Have you ever seen a climber try to catch a chicken at a rodeo?

Adriana fell in love with Joe's Valley the same way I did, so much so that she moved there permanently. She was working for the county while restoring an old property in town and one day forwarded me a request for proposals from Emery County for a social media strategist. The Swell Utah wanted a slice of our state's booming tourism pie—and if anyone was going to put my most beloved desert on the map, it was going to be me.

The San Rafael Swell is seventy-five by forty miles of sandstone, limestone, and shale, home to tricky slot canyons, towering mesas, expansive badlands, ancient geological formations, and all of my favorite dirt roads. Most folks pass right through on their way from Salt Lake City to Moab, but the Swell's landscapes rival any Utah national park. It is my favorite place on earth.

I created an eleven-page pitch document that contained all the catch phrases and technical strategy jargon to prove my chops as a social media professional, but the real sell had to be delivered verbally. It required the kind of honesty reserved for direct conversations. I explained to the commission and travel board that I was the right tool for this job because I was going to "respectfully, make us the antidote to Moab's brand of tourism."

Everyone in the room chuckled, and one colorful local clapped his knee as he cackled.

Moab, you see, is a classic tale of when too much of a good thing becomes a bad thing. If you've visited in the last five years, you've found it hard to believe that Moab was once a sleepy little rural blip on the desert radar. Today, its crown-jewel national park, Arches, is so crowded it requires reservations just to drive your car through the gates. The tourism boom has affected the local community too, with housing shortages and major residential conflicts with off-highway vehicle (OHV) users. Traffic backs up a mile down the highway before town even begins.

My vision for the Swell's successful recreation economy centered around sustainability. We didn't want tourists coming to Emery County, we wanted ambassadors; folks who came to visit and left in love. The tagline was, "When you visit the Swell, you're not just a visitor here—you're part of this place."

I wrapped up my presentation by saying, "As a longtime advocate of Emery County's unparalleled outdoor opportunities, unique history, and warm community ethos, it would be the project of a lifetime to have a hand in developing the next chapter of tourism in this special place."

The text message I received after my pitch ended said, "I think you got it."

Before I was officially hired, I had to tiptoe through a gauntlet of rural politics to prove my trustworthiness. I acknowledged that while I might be a vegetarian (read: registered Democrat), I would never say no to local legend Ungerman Meats in Emery County—and that I would respect the political leanings of county leadership while working to serve them. My casual reference to the county's most adored butcher sealed the deal.

We began our work with intensely long stake-holder engagement sessions. I mostly listened, but occasionally offered my perspective as a longtime vis-itor. Most folks were eager to improve the economy through recreation but skeptical that outsiders would be invited to share the space. Though they were set-tlers themselves, many of the families there had been holding the Swell's landscapes sacred for generations. While the need to diversify the community away from oil and gas was palpable, so was the fear of tarnishing the sanctity of an even more precious resource: the landscape itself.

This place was special to me for its obscurity and solitude. The Swell is "my" spot. Would implement-ing a successful branding strategy catapult my favorite campsites from off-the-beaten-path to always-occupied? What if my antidote plan didn't work and I unleashed a beast of visibility and visitation beyond Emery County's capacity?

My first act as strategist for Emery County's digital tourism voice was to attack and destroy my least favorite brand language displayed prominently across the top of TheSwellUtah.com. In big, loopy font it declared, "The playground is open!"

Describing the Swell as a playground did not pass the vibe check.

If we wanted people to leave this place with a sense of ambassadorship and pride, we needed to lead with messaging that sent folks down that path. Hoping to inject more local flavor into the process too, we brainstormed a list of Emery County-isms like "'ppreciate ya," and "keep it swell." Combined, the two phrases created the perfect tag line for this unique place: *'ppreciate ya keeping it swell!*

In the beginning, I envisioned my time with Emery County would take us to a baseline of social media engagement that might even make us a little uncomfortable. After the first few stakeholder meetings, I understood that visibility wasn't the goal here. I stopped scheming my usual tactics for quick organic growth on social platforms and started focusing on daydreams about how my ideal visitor to the Swell would act when they came to visit the sandstone heaven—and how I could play the role of shepherding them toward that type of behavior.

By the time I completed my work with Emery County, the project had expanded from a six-month contract to an eighteen-month endeavor—and I wouldn't have shortened it for a second. In March 2020, I wrapped up training with the woman who would take the reins on Emery County's social media. She was a local woman with marketing experience, public lands management experience, and a husband who worked for the Bureau of Land Management. We drove out to Swaysey's Beach in Green River and practiced content creation. She took the videos we shot and created a storytelling series about how to visit the Swell and practice Leave No Trace ethics. My job was done.

When I pulled back into Green River that day, I checked into my hotel and was greeted by the owner.

"Katie Boué?" He asked with a wide grin.

"That's me!" I responded, a little nervous after a season of internet hate.

"I'm Travis. We've been on many calls together while you've been working with the tourism board. I just want you to know, you're welcome here in Emery County any time."

I thanked him, took my key to my room, and wept.

Today, the banner spread across the top of the Swell's website has been updated. It now reads: "Keep it Swell."

Social media engagement is still modest for the county, but when the boom of tourism does come—and it will—the Swell will be ready to lead visitors toward a different type of adventure. Through the language and tourism ethos we created for Emery County, a legacy of stewardship, curiosity, and respect will live on as more and more people discover the magnificence of my favorite place on earth.

SNARK AND CONSEQUENCES

The darkest part of being a public figure, particularly in the advocacy space, is the lateral violence inflicted from within our own movements.

For years, I have been one of many "influencers" targeted by a snark forum about outdoor Instagrammers. It was an insidious corner of the internet that bred harmful accusations, invasions of privacy, violent harassment, and the general degradation of human decency.

The ferociously inhumane things I saw about myself and my loved ones in those threads became addictive to read. As a public-facing person, I felt obligated to listen to what "the public" had to say about me—both the good and the bad. I believed the hatred I received on these threads was part of being a visible advocate, so I subjected myself to it every day. Reading the opinions of folks who hated me as a hobby led me to develop suicidal ideations, amplified by the depression and anxiety that had blossomed during the pandemic.

When my home address was doxed and passed around to malicious strangers, I became agoraphobic. I developed an irrational fear of going to trailheads, being in public, even eating at my favorite neighborhood restaurant after my regular server, who once told me she was a fan, ended up in the threads discussing my tipping. I no longer felt safe, either in the digital or real worlds.

There were a handful of legitimate, important criticisms buried within those comments, but the legitimacy of anonymous feedback was vastly overshadowed by the toxic, harmful noise powering those threads. Folks copied photos of my partner and drew obscenities on them. They discussed how much money they'd be willing to pay to punch him in the face. A Discord channel was created in obsession with his feet. They watched everything I posted on online, speculated on my life based on my Pinterest activity, and even stalked my father's LinkedIn profile trying to ascertain whether I have a trust fund (I don't). Yet they'd express shock when their actions were described as harassment. Some even claimed the hate was in the name of accountability—though others quickly admitted it was really all just for sport.

The tipping point for me came in May 2022, when a particularly nasty woman who thought she was anonymous—let's call her Flynn Rosenbaum—decided to post a link speculating about my "next venture." Nosy as always, I clicked the hyperlinked blue text, and it opened up a screenshot. The image was taken from one of my close personal friend M's new consulting business.

This friend, M, is not an influencer, nor a public figure in any way. Her Instagram account is private and only has a small following of her friends, family, and colleagues. This meant that Flynn, in her pursuit of hating Katie-Boué-the-Advocacy-Influencer as a hobby, somehow weaseled her way into a private person's space with the intent to weaponize that access to further harass and degrade me.

As she mocked the new business venture and insulted M's professional logo design thinking I had created it, I felt my blood boiling. For two years I had endured Flynn's bullying specifically targeted at me and my partner, people who chose to become public entities. But this attack on an innocent bystander crossed the line from appropriate online behavior into an unacceptable and intolerable violation.

So, reader, I finally decided to do something about it.

You see, about six months prior I received an anonymous spreadsheet containing a list of names, addresses, employers, and contact information. It was delivered by a woman, we call her Selena, who had heard about the snark threads, decided it was wrong, and used it as practice for her hobby of tracking white supremacists online. Using her sleuthing skills, she was able to identify most of the key participants in this digital harassment, including moderators.

This meant I knew who the anonymous user really was. So, I reached out to Flynn directly from my Instagram to her personal account.

"Hi Flynn!" I wrote. "It was super inappropriate for you to post screenshots from my friend's private page, she's a private person, not an influencer. I understand that you've let me and my partner live rent free in your head every day for the eight months since I've last posted on Instagram, but don't be gross dude. Thanks!"

I hit send.

Flynn blocked me moments after I sent the message, so I know she received it, but I may never know what she replied or if she said anything at all. I thought I saw a flash of a response, but I'll likely never see it. Hours later, her Reddit account was gone. The horrible things she spent years publishing about me online remain. She has since made a few new Reddit accounts under different names.

I breathed a sigh of relief after the small victory, but my hands were still trembling with rage and anxiety.

Many months earlier, Flynn posted that she was embarrassed to admit to her husband that she participated in those awful threads. I wonder how she felt when she realized that I knew every single horrific thing that came out of her mouth—and I also knew her address, her employer's contact information, and who all our mutual friends are.

She wasn't, and had never been, anonymous. She had been herself, Flynn Rosenbaum, a small-town elections worker who spent her free time cyberbullying strangers for years of her adult life. Every single word she wrote under the false guise of anonymity was in fact tied to her real identity the whole time. Anonymity no longer exists, and we must act accordingly.

When I messaged Flynn, I also wrote directly to one of the identified moderators of the group, asking her to remove the inappropriate posts. She too blocked me before I could see any response, but the threads were made private immediately.

The first time I clicked the forums in anticipation of another round of torturing myself by reading snark and was halted by a privacy wall, I was freed. The week I announced the publishing of this book, the threads were made public again—but by then I had broken the habit of reading it, and I have not visited since.

As Selena continued her pursuit against cyberbullying, the spreadsheet of once-anonymous snarkers grew in size each week, and multiple lawyers offered consultation on how best to proceed. There were discussions of launching defamation lawsuits, printing out copies of snarkers' worst comments and mailing it to their employers, even a fantasy of reading out their real identities on a live stream while drinking margaritas. But how does one seek justice in a situation muddled with ugliness at every angle?

The petty Miami villain in me wanted so badly to turn this all into a real-life telenovela complete with revenge and chaos. To seek revenge against Mae, the Washington woman who befriended me over four years of exchanging gardening messages, all while secretly leading the worst attacks against me on multiple Reddit accounts. To publicly humiliate those who found it entertaining to publicly humiliate me.

Unfortunately for my occasional fantasies about publishing the entire list of snarkers online and watching the internet burn, the cycle of cyber abuse needed to stop—and the buck was now with me.

Hateful spaces are waiting at every corner of the internet to smother us with toxicity and the addiction of drama. It's a sticky trap, a trick hoping we'll slip up and forget what mutual respect and minding our business looks like. Hate is enticing and entertaining—but hate is not helpful.

Behind every infographic, email newsletter, brand account, and direct message you'll find a human being.

When you reach out on social media, you aren't contacting an ethereal brand entity or celebrity, you're reaching a regular person—likely underpaid, undervalued, and often incapable of moving messages up the chain quite as effectively you'd hope. Online, treating humans with kindness means leading social media interactions with kindness. Remove the screens, the keyboards, the fiber internet cables, and all you have left are humans.

When we consider our online behavior, we must do so as if these were real-world interactions. It's as simple as: Would I say that to someone's face? Would I stand up in a crowded room and say that? The rules of human decency are not suspended when we engage through keyboards and touchscreens. There's important power in anger, yes, but there's no purpose in just being an asshole. Not to be confused with toxic positivity, what we need is an unwavering, universal commitment to upholding the bounds of mutual decency, even in challenging circumstances.

Digging into the identities of these people—nearly all women—I discovered depths of lifelong trauma, injury and illness, severe mental crises, devastating loss, addiction and abuse, breakups, financial struggles, and familial conflict. I scrolled through their digital personas, learning that the snark that had traumatized me had nothing to do with me. It wasn't about me. My Instagram personality was merely the vessel for strangers projecting their own unhealed pain into the ether of the internet. The hate against me had nothing to do with advocacy.

Even as I sat on the receiving end of abuse disguised as accountability by these women online, it was my responsibility to choose my decency instead of my suffering. Hurt people hurt people, and the pain of others is frankly none of my business—so there's nothing I can do but forgive.

So, that's what I did.

Cease and desist paperwork was sent by lawyers, and the spreadsheet was delivered to everyone else harassed by the snarkers. Eventually, my friend Brianna Madia called out our harassers, and along with Selena we deleted and dismantled all of the threads—but that's a story for another time. I know I'll never erase online hate about me forever, but I also know it's not about me anymore.

It never was.

I'll never read hate threads about me again—I'm too healed to go back there. I respect myself and my advocacy too much to subject myself to frivolous suffering. We've all got way too much work to do to spend our time reading the comment section.

It's a powerful thing to log off the drama and get back to work.

Drink Water, Mind Your Business

My dear friend Caroline Gleich once said, "The most powerful form of activism comes from a place of joy and happiness." I think anger deserves accolades as a powerful force too, but joy is what really fuels us. Even in the deepest pits of anger, it is the promise of a joyous future that we fight for.

Advocacy will always include challenges, hardships, frustration, and pain. An allegiance to joy isn't a rejection of the hard parts, it is a dedication to the hope we hold through it all. Our job isn't to fix every problem and carry the weight of every crisis—all we can do is stay hydrated and mind our own business. Build a set of beliefs, and mind them.

I believe that the planet can be saved.

I believe that there is enough "outdoors" for all.

I believe outdoorists can be good neighbors.

I believe that's all worth advocating for.

As you walk the path toward joyous advocacy, you will inevitably encounter people who are resolute in their commitment to misunderstand you. Don't waste your energy on them. You have no obligation to engage with someone who has no intent of listening or reciprocating respect. Spending your time pinging back and forth with someone hateful isn't advocacy. It's torture. Build a circle of friends and colleagues you can reciprocate accountability with. People who adore and uplift you, but also challenge and communicate directly with you. Ask for hard feedback from people who care about you, and take it seriously.

Drink your water, and mind your business.

Keep your friends accountable in doing the same.

TIDINGS TO LEAVE YOU WITH

As you go forth and advocate, take the following tidings and put them into practice. Throughout each of these advocate's exercises is a throughline of always finding balance, taking care, minding your business, and hydrating.

Be wary of experts.

Have a critical eye toward what content you're consuming and by whom. Check your sources. And for goodness' sake, beware of becoming a so-called expert. That's the trap right there.

During the decade I have spent working in the outdoor industry as a social media communications professional, I've built up expertise in my fields. Expertise in one field, however, does not mean all-mighty expertise on all things. That is healthy. Seek people who know what they are good at and stick with it.

Understand that sharing on social media is not a demonstration of expertise. Anyone can hop online and speak as if they are an authority. This is how misinformation spreads. Understand the lived experiences, education levels, and field expertise of an individual. Set your expectations of the people you follow accordingly.

Stop taking yourself so seriously all the time. Be weird, make art.

If social media is all performative, then let's make a beautiful performance out of it, right?

The day my digital presence started going downhill was the day I started confusing the canvas for the pearly gates. Social media and our digital identities are not places for us to provide our worthiness as righteous martyrs "doing the work" and suffering for the cause—it's a container for snippets about what we're up to, what we're inspired by, and the world we're occupying IRL.

What I'm trying to say is: stop taking Instagram so damn seriously. We aren't going to fix the influx of visitation on delicate ecosystems in the comment section. No solutions for permitting or pollution will be magically born from my Instagram story posted on a Tuesday afternoon. Our social media pages are powerful, yes, but they are what they are: and that's just an interactive content space.

The timelines of when I started receiving real, targeted harassment online and when I started taking myself too seriously match up perfectly. People can smell bullshit from a mile away, and I was a steaming pile of it. Unhappy, scared, spiritually burnt to a crisp, I kept up the performance long after I should have hung up my paintbrush. Abusive or not, even strangers on the internet could see straight through me. Nobody likes an artist who is faking it.

The digital space isn't real life, so use it for what it can be at its best: a playground for curiosity. I used to spend days toiling over the perfect picture-of-me-on-a-mountain paired with an emotionally evocative caption about federal public lands legislation. With respect to my former self, that was not the best use of my time. What if I had spent that time writing a compelling testimony or public comment to submit to my lawmakers—which I could then still use as an Instagram caption? A small shift of intention, but a bigger step toward doing good. My time is better spent not just making it about me. So is yours.

My ultimate criticism of the outdoor advocacy influencer scene is this: (for now) it seems we are unable to create successful advocacy movements that aren't centered upon our own egos and individual clout. Period. Whether well-intentioned or not, using advocacy as an entire personality and career creates unavoidable errors of ego and misspent energy. Remember what I said before about being wary of the experts?

What I'm trying to say when I say "be weird and make art" is: be your whole human self. You are an entire, complex, strange, wonderful being, and you deserve to exist and be seen in your fullest form. When we limit ourselves to a digital-activist persona, we often inhibit all the best parts of ourselves. If you're stoked about mountain bikes and trail access, you can also be a big fan of jazz bars and estate sale shopping.

It's exhausting being righteous all the time. We should be online less, yes, but when we do show up, we must do so with a spirit of curiosity and fun again. Post that picture of your latest backyard gardening success, channel your advocacy urgings into an Instagram Live fundraiser bake sale, get nerdy and do a series profiling all the different types of bugs you can identify at your local crag.

Be the weird you wish to see in the world.

Maybe you'll start seeing more of it mirrored back at you.

Maybe we'll all start to be more weird (read: happy) together.

Keep it human.

What happened to the humanity of it all? What happened to brotherly love, neighborly love? When exactly did we go from love thy neighbor to cancel thy neighbor over hot takes on the internet? How do we reel it back in now? Can we?

The solution for me came in shrinking my circles. After the Great Burnout of 2020, I radically popped the giant bubble in which I had resided for the past decade and relocated my tattered soul to a much smaller, much safer space that could only be held by the confines of Utah's Wasatch Front. In shrinking my circles from the ethereal "everyone" of the internet to a hard focus on my neighbors IRL, I opened myself back up to the idea of our collective humanity.

We're all just humans, living next to each other.

Just a bunch of bone piles, scraping our skin sacks along sandstone walls, bobbing our bony bodies along rapids, baking our salty epidermises in the sun. To humanize ourselves and others again is a disarming experience. We're all somebody, and we're all just nobodies. It'd be tempting to pick a fight between the dichotomies of whether we're all special or all just the same—but I think our imagination is better spent celebrating how cool it is to be both.

Humans share 99.9 percent of our DNA with one another. We're all just a bunch of soft-skinned souls wandering around the earth at the same time, together. Be a human, keep it human.

You have two hands to hold two different things at the same time. Use that skill often.

Become a "yes and," type of thinker. It's a basic premise of improv I learned decades ago as a young thespian in Troupe 1298 at Palmetto Senior High School in Miami, Florida. A successful improviser always operates on a philosophy of "yes, and," never a "no, but." You always build on the bit.

You've got a big, curious brain. Use it to expand, not contract. The problems and conversations facing our communities these days aren't the type of discussions easily folded up like a dinner napkin and tucked into a decorative ring. We're addressing issues that deserve to occupy the space of the entire dinner party.

We're pursuing big, audacious issues and ideas—don't you think that deserves to occupy big, complex spaces? If you have two hands available, why would you try to fit the entirety of a topic as large as, say, reconciling the violent history of public lands into a single handful? I think we'd agree it deserves at least two.

So, be an invitation to get bigger. Steward opportunity.

If something makes you uncomfortable, hold it with one hand. Use your other limb to make space to carry another possibility. Maybe a rebuttal or opposing argument or radically imagined alternate scenario.

We all need to make a little bit more room—on the trails, in our digital spaces, within our communities—and creating space in your critical thinking to hold with both hands empowers you to be a better neighbor, strategist, and solutions builder.

We've lost our ability to juggle. We've lost the belief in ourselves that we're capable of holding more than one thing at a time. We've become so precious with the one thing we are holding that we've grown to distrust the possibility of having strength to carry anything else too. As if the basic act of opening our minds to a new perspective could threaten our grip on the one thing we still have in this tumultuous, untrustworthy world.

Friends, we can hold so much. Certainly much more than the one white-knuckled grip we have on our Sole Remaining Original Thought That Is the Hill We Are Willing to Die On. Believe in yourself. Believe in your beliefs. Opening yourself up to a conversation in good faith with a neighbor whose bridge you have no intentions of burning is not a threat to your Original Thought's existence. It might be an affirmation, or a challenge, but those are both good for you.

Keep it sustainable—not just for the planet, but for yourself.

In 2020, we learned the term "burn out," and we embodied it from head to toe. As we emerge into new normals and open our eyes to doing things differently, we have the opportunity to create more sustainable lifestyles.

I'm not talking about switching to bar shampoo or driving an electric vehicle—I mean personal sustainability. The ability to sustain yourself: a human, a life force, a nutrient-needing body, a rest-requiring soul.

We have learned to decipher our bodies' cries of "please, it's just too much" as a sign that we should push onward even harder. We've started to believe that if we suffer more, we're "doing the work" better. We've been taught that pain is productivity.

What if rest, giving, receiving, and joy were what we viewed to be productive?

Wouldn't it be easier to continually show up and give your all if it wasn't so damn exhausting and harmful to your soul?

Learning personal sustainability is a radical act of self-care and resilience building for advocates. It felt selfish at first, to start prioritizing myself in the spirit of "pull the mask down over your face before helping other passengers." Not only to break up with the self-imposed importance of having to suffer all the time, but to truly slow down and ask what it would look like to set up a life that allowed me to continue toward advocacy every single day without burning out.

Here's a short list of changes and small realizations I had that shifted my personal sustainability as an advocate:

1. **You can't be everything all the time.** In the spirit of embracing that we are all just finite skin sacks floating through space and time, we must be realistic about what our fleshy hands and brains are capable of. One simply cannot be an expert on all current affairs, constantly sharing relevant and accurate resources, responding to an infinite volume of engagement, and still functioning as a human being—trust me, I've tried. Forcing expertise and overextending your energy is disrespectful both to yourself and to your community.

2. **Examine the sense of urgency, and remember that it's likely not that urgent.** One of the tricks of the internet is a relentless sense of urgency. Everything is immediate, breaking, critical, and top priority— but if everything is, then eventually nothing is. The fast-paced nature of social media suggested that inflammatory language become part of the process

of capturing a user's attention. Urgent matters do exist, of course, but the key is to learn how to identify what demands your energy now and what can wait. Ultimately, this is part of the process of unlearning our self-imposed importance.

3. **Related: You don't have to respond right away.** This has layers. When I quit my job with OIA, I had some reckoning to do with my inbox habits. I had come into the habit of believing emails required immediate response, no matter if it was dinnertime, a weekend, or while I had an Out of Office up. If I didn't reply right away, the world would surely fall apart. Right? Wrong. Similarly, with responding immediately on social media to real-world tragedies seemed critical. We have built a culture that expects us to see horror, heartache, injustice, violence, extremism, pain—and instantly react with a hot take on Twitter or aesthetic graphic with a GoFundMe link in our Instagram stories. That is not a natural human way to process and grieve. These tiny fleshy bodies and brains were not meant to handle the weight of an entire dying planet. Take a night. Take two.

4. **You actually just don't need to be online, at all.** You can log off at any time and still be a good advocate. That statement will get me some rabid hate mail, but I

stand by it regardless. Your contributions to advocacy, your ability to empathize, your worthiness as a human being and advocate has nothing to do with whether you can stomach doom scrolling. When the internet starts to feel like a knot in the pit of your stomach, put your phone down. I "logged off" for eight months and the world kept turning. This is not an invitation to ignore what's happening around us, rather it's a reminder that sometimes we can choose to experience it offline.

An upside of taking the spooky step of logging off when it feels bad is that you'll find yourself ready to log back on with better intentions. When we give ourselves permission to leave the digital space when it becomes harmful, to wait on responding to that email until the morning, we are setting better boundaries with what we consume and share online.

CHECK OUT WITH A CHECK-IN

I've become a fan of the self check-in. Accountability is best served by those who know us intimately—so who better to call you out and call you in than yourself? Here are a few questions to ask yourself when checking in:

- How are you spending your time and energy?
- Who are you voting into office?
- What is your advocacy support circle?
- What brings you joy in advocacy? What brings you joy in life?
- How are you spending your money?
- How are you giving back to your community?
- How are you organizing your circle?
- What would collective liberation look like for you?

When I ask myself those questions now, my answers are radically different than they would have been one year ago. I'm spending my time and energy engaging in my local community, tending to my gardens, making goods with my hands, and enjoying more quality time with a supportive circle of co-conspirators. Without the pressure of representing every major public lands issue across the country, I'm paying closer attention to Salt Lake City politics than ever. By shrinking my circle to better fit my capabilities as just-one-human, I made myself available to being better educated and better armed for advocacy action.

In taking an extended break from the ways I used to embody advocacy, I allowed myself permission to use my creative imagination to build an entirely new version of what organizing and giving back can look like. I crafted a Little Free Seed library and have been distributing vegetables and wildflower seed packets to budding gardeners at the markets where I vend. Months later, I spent six weeks meeting after-hours at a café to help local bakers organize an abortion fund bake sale. I donated floral arrangements and volunteered my expertise to build a communications toolkit for the fundraiser. On the day of the bake sale, there was a line of people around the building waiting to get in. We sold out completely within two hours.

For the first time in a long time, I have the energy and shits-to-give about new advocacy issues. I'm eager to learn more about how I can help efforts to Save the Great Salt Lake. I've exchanged numbers with new friends working on solutions for community problems. I'm looking forward to logging back onto my main Instagram account after hitting send on this manuscript. I'm ready to come back. I'm ready to be an active advocate again.

Though once you become an advocate, you're in for life.

We all have a lifelong responsibility to advocate for the outdoors.

As users of natural resources, we have an obligation to take care of the earth because we love these places, because we need them, and because they cannot take care of themselves. When we protect the land, we are protecting people and communities too.

There's no official certification, set of qualifications, or test to become an outdoor advocate. Congrats, you're an outdoor advocate now, forever. Go forth and uphold your responsibility with joy, tenacity, and curiosity.

Take one action today and take another tomorrow, and keep taking action until outdoor advocacy becomes an integrative habit in your life.

Being an advocate is not just something you do; let it become a vibrant part of who you are.

ACKNOWLEDGMENTS

To my parents and sister, without whom I wouldn't exist. Every dream I've ever had was always within my reach because of you. For the earthworm rings, rainstorm hikes in the desert, and hands that have become too soft from laughing. *Mi familia es el mejor regalo de mi vida.*

To Amelia Howe, my life and limb—and Dory, Madeline, and Grace. To women supporting women. *Por que no los dos.*

To Spaghetti and Brody, for making sure I didn't stay in my writing cave too long without taking a break to sit with the bees. I love you forever.

To Jenn Brunson, who hired me for my first outdoor industry job, and fiercely believed in my voice. 11:11 forever.

To Vivian Wang, without whom I would cease to function professionally. You are a sister to me, and I'm proud of you.

To my editor and Fulcrum Publishing for taking a chance on this, and for giving me my voice back. I thought it might be better to be quiet, but you reminded me I exist to be loud.

To Rachelle, for putting air into my lungs when I forgot to breathe.

To Selena, you sneaky bitch. For giving me my peace back.

To my gardens, for keeping me in the light through the darkness.

To the earth, how did we get so damn lucky to exist here, now? I promise we'll take better care.

GLOSSARY

ambassador program. A (typically) ongoing marketing campaign by a brand, where a group of individuals are selected as brand representatives in exchange for payment, gear, trips, and so on.

community care. The concept that in order to sustain thriving and engaged advocacy communities, we must take care of each other and ourselves.

crytobiotic soil. Per the National Park Service, "Cryptobiotic soil crusts are created by living organisms such as algae, cyanobacteria, and fungi. The bacteria within the soil release a gelatinous material that binds soil particles together in a dense matrix. The result is a hardened surface layer made up of both living organisms and inorganic soil matter" (source: https://www.nps.gov/glca/learn/nature/soils.htm).

digital advocacy. According to the University of Kansas, "Digital advocacy is the use of digital technology to contact, inform, and mobilize a group of concerned people around an issue or cause" (source: https://ctb.ku.edu/en/table-of-contents/advocacy/direct-action/electronic-advocacy/main).

direct action protest. The *Activist Handbook* says, "Direct action is a form of protest in which those taking part seek to achieve their goals through direct, often physical, action, rather than through negotiation or discussion" (source: https://activisthandbook.org/tactics/direct-action).

dispersed camping. According to the US Forest Service, "The term used for camping anywhere in the National Forest outside of a designated campground. Dispersed camping [means] no services and there are few, if any, facilities. There are extra responsibilities and skills that are necessary for dispersed camping." This concept also applies to other public land management agencies where allowed (source: https://www.fs.usda.gov/Internet/FSE_DOCUMENTS/fseprd908212.pdf).

grass roots/grassroots. Typically used in political or organizational contexts, grass roots means returning to the most basic or local level of individual engagement of an advocacy community.

Leave No Trace. A set of outdoor ethics created by the Leave No Trace Center of Outdoor Ethics promoting conservation in the outdoors, with principles ranging from respecting wildlife to properly disposing of waste.

lobbying. Seeking to influence someone—typically a politician, public figure, or other stakeholder—on a particular issue.

outdoor advocacy. The idea of building community and action around issues affecting the outdoor space, particularly as it pertains to land conservation, recreation, and other topics related to the natural world.

Outdoor Advocacy Project (OAP). A communications consulting group and community media organization founded in 2019.

outdoor industry. According to Colorado College, "The outdoor industry is comprised of companies that promote exploration in the outdoors and/or distribute/manufacture outdoor apparel, gear, equipment, accessories, and footwear for outdoor activities" (source: https://www.coloradocollege.edu/offices/careercenter/explore-careers/sciences/outdoor-industry.html#:~:text=The%20outdoor%20industry%20is%20comprised,and%20footwear%20for%20outdoor%20activities).

Outdoor Industry Association (OIA). A nonprofit trade association for companies like retailers, manufacturers, and media agencies in the active outdoor recreation business.

outdoorist. A term coined by the marketing team at Outdoor Industry Association to identify individuals within the outdoor recreation community.

Outdoor Recreation Jobs and Economic Impact Act. A piece of 2016 federal legislation that "directs the Department of Commerce to enter into a joint memorandum with the Department of Agriculture and the Department of the Interior to conduct, acting through the Bureau of Economic Analysis, to assess and analyze the outdoor recreation economy of the United States and the effects attributable to it on the overall U.S. economy." (source: https://www.congress.gov/bill/114th-congress/senate-bill/2219).

public lands. "In the broadest sense, [public lands are] areas of land that are open to the public and managed by the government. You can think of it as land you own (and share with everyone else in the United States)" (source: https://www.rei.com/blog/hike/your-guide-to-understanding-public-lands).

reciprocity with the land. Giving back to the environment in understanding that we receive so much, like stress relief and joy, from our relationship with the outdoors.

recreate responsibly. A series of outdoor guidelines and a nationwide collaborative campaign developed by the Recreate Responsibly Coalition to inspire an outdoor culture of stewards to our communities, land, and water.

stewardship. Merriam-Webster defines stewardship as "the conducting, supervising, or managing of something. especially: the careful and responsible management of something entrusted to one's care."

sustainability. UCLA says, "Sustainability is the balance between the environment, equity, and economy" (source: https://www.sustain.ucla.edu/what-is-sustainability/).

sustainable activism. The idea that our activism is a resource we must be careful not to deplete without the ability to replenish it, ensuring that we are able to continue our advocacy long-term without burning out.

Tread Lightly. A nonprofit organization with a mission to promote responsible recreation through stewardship, education, and communication.

ABOUT THE AUTHOR

Katie Boué believes we all belong in nature and have a responsibility to be stewards of the earth and communities around us. She is a Cuban-American outdoor advocate, gardener, adventurer, and budding bird-watcher. When she's not out communing with nature, Katie spends time as an award-winning marketing strategist, writer, and content creator.

Katie was the recipient of the Hero for the Planet Award at the Outdoor Media Summit in 2017 and has been featured in publications like *Outside*, *Lonely Planet*, *SELF*, the *Denver Post*, and many others. She earned her communications chops over ten years of social media marketing and digital organizing in the outdoor industry before burning out and finding balance through gardening and backyard stewardship.

Born in Queens, New York, and raised in Miami, Florida, Katie currently lives on a half-acre urban homestead in Salt Lake City, Utah, with her dog, Spaghetti, and partner, Brody Leven.